Electronic Health Records - Issues and Challenges in Healthcare Systems

Edited by Venkata Krishna Parimala

Published in London, United Kingdom

Electronic Health Records - Issues and Challenges in Healthcare Systems
http://dx.doi.org/10.5772/intechopen.1003290
Edited by Venkata Krishna Parimala

Contributors
Adrian Victor Vevera, Ajna Fetić, António Dionísio, Bob Chaput, Carmen Elena Cîrnu, Christos N. Schizas, Constantinos S. Pattichis, Elena-Anca Paraschiv, Filipa Martinho, Maria José Sousa, Nevzudin Buzađija, Pedro Ramos, Ricardo Marcão, Ritviksiddha Penchala, Sakshi Porwal, Stéphanie Monteiro, Theodoros Solomou, Vasco Santos, Venkata Krishna Parimala

First published in London, United Kingdom, 2024 by IntechOpen
IntechOpen is the global imprint of INTECHOPEN LIMITED, registered in England and Wales, registration number: 11086078, 167-169 Great Portland Street, London, W1W 5PF, United Kingdom

For EU product safety concerns: IN TECH d.o.o., Prolaz Marije Krucifikse Kozulić 3, 51000 Rijeka, Croatia, info@intechopen.com or visit our website at intechopen.com.

British Library Cataloguing-in-Publication Data
A catalogue record for this book is available from the British Library

Electronic Health Records - Issues and Challenges in Healthcare Systems
Edited by Venkata Krishna Parimala
p. cm.

This title is part of the Biomedical Engineering Book Series, Volume 28
Topic: Bioinformatics and Medical Informatics
Series Editor: Robert Koprowski
Topic Editor: Slawomir Wilczynski

Print ISBN 978-0-85014-945-6
Online ISBN 978-0-85014-944-9
eBook (PDF) ISBN 978-0-85014-946-3
ISSN 2631-5343

If disposing of this product, please recycle the paper responsibly.

IntechOpen Book Series

Biomedical Engineering

Volume 28

Aims and Scope of the Series

Biomedical Engineering is one of the fastest-growing interdisciplinary branches of science and industry. The combination of electronics and computer science with biology and medicine has improved patient diagnosis, reduced rehabilitation time, and helped to facilitate a better quality of life. Nowadays, all medical imaging devices, medical instruments, or new laboratory techniques result from the cooperation of specialists in various fields. The series of Biomedical Engineering books covers such areas of knowledge as chemistry, physics, electronics, medicine, and biology. This series is intended for doctors, engineers, and scientists involved in biomedical engineering or those wanting to start working in this field.

Meet the Series Editor

Robert Koprowski, MD (1997), Ph.D. (2003), Habilitation (2015), is an employee of the University of Silesia, Poland, Institute of Computer Science, Department of Biomedical Computer Systems. For 20 years, he has studied the analysis and processing of biomedical images, emphasizing the full automation of measurement for a large inter-individual variability of patients. Dr. Koprowski has authored more than a hundred research papers with dozens in impact factor (IF) journals and has authored or co-authored six books. Additionally, he is the author of several national and international patents in the field of biomedical devices and imaging. Since 2011, he has been a reviewer of grants and projects (including EU projects) in biomedical engineering.

Meet the Volume Editor

Dr. Venkata Krishna Parimala is currently a Professor of Computer Science and the Director of University Rankings at Sri Padmavati Mahila Visvavidyalayam, Tirupati, India. He received his BTech in Electronics and Communication Engineering from Sri Venkateswara University, Tirupati, India; an MTech in Computer Science and Engineering from Regional Engineering College (REC), Calicut, India; and a Ph.D. from VIT University, Vellore, India. Dr. Krishna has several years of experience working in academia, research, teaching, consultancy, academic administration, and project management roles. His current research interests include mobile and wireless systems, data science, explainable artificial intelligence (XAI), machine learning, and cloud computing. He is the recipient of several academic and research awards, such as the Cognizant Best Faculty Award for 2009–2010 and the 2021 Best Researcher Award from Sri Padmavati Mahila University. Recently, Prof. Krishna was listed among the top 2% of scientists in the world by Stanford University, USA, in 2021. He has authored over 250 research papers in various national and international journals and conferences. He has delivered several keynote addresses and chaired sessions in reputed conferences. He is a senior member of several professional societies, such as the Institute of Electrical and Electronics Engineers (IEEE), Association for Computing Machinery (ACM), Computer Society of India (CSI), Institution of Engineers (India) (IE(I)), and others.

Contents

Preface

The rapid digital transformation of healthcare has placed Electronic Health Records (EHRs) at the center of modern clinical practice, public health strategies, and patient-centered care. As EHR systems evolve in response to growing demands for data-driven decision-making, interoperability, and personalized medicine, concerns regarding data security, system integrity, and usability have emerged. This book, *Electronic Health Records – Issues and Challenges in Healthcare Systems*, presents a timely and multi-domain exploration of the current landscape of EHR adoption, with particular focus on cybersecurity, artificial intelligence (AI), mobile health innovations, and systemic optimization of healthcare delivery.

The book is organized into three sections that address the issues and advancements surrounding EHR systems.

The first section, *Cybersecurity and Risk Management in EHR Systems*, addresses the essential need to safeguard patient data and ensure system resilience. Chapter 1, *Assuring Patient Privacy and Safety through Robust Enterprise Cyber Risk Management (ECRM),* outlines the critical components of enterprise risk management in healthcare, emphasizing frameworks that support compliance, data protection, and organizational continuity. Chapter 2, *Integrating Artificial Intelligence and Cybersecurity in Electronic Health Records: Addressing Challenges and Optimizing Healthcare Systems*, investigates the dual roles of AI in enhancing system efficiency and introducing new vectors of vulnerability, offering strategies to strike a balance between innovation and security. Chapter 3, *Perspective Chapter: Securing the Future of Digital Healthcare with EHR Systems for Modern Cyber Threats*, provides a forward-looking perspective on securing EHR infrastructure against sophisticated cyber threats through a combination of policy, technology, and training initiatives.

The second section, *Advances in Mobile and AI-Driven Health Technologies*, explores integrating cutting-edge computational methods and mobile platforms to enhance patient outcomes and operational efficiency. Chapter 4, *Emerging Mobile Health Systems and Services*, discusses the transformative potential of mobile health (mHealth) tools in bridging healthcare accessibility gaps, especially in remote or underserved regions. Chapter 5, *Evaluation of U-Net Algorithm Precision for Segmentation and Classification of Breast and Lung Images on CBIS-DDSM and LIDC-IDRI Datasets*, presents an evaluation of AI-powered image segmentation techniques, showcasing how convolutional neural networks can assist radiologists and clinicians in early and accurate diagnosis using real-world medical imaging datasets.

The third section, *Innovation, Optimization, and Strategic Management in Healthcare Systems*, focuses on the broader implications of digital health integration on system performance and economic sustainability. Chapter 6, *Optimizing Healthcare Delivery: Innovations and Economic Strategies in Medical Device Management,* addresses how strategic procurement, device lifecycle management, and digital monitoring systems can enhance hospital efficiency while ensuring patient safety and regulatory compliance.

Finally, the contributions in this book reflect a diverse and interdisciplinary approach to EHR research, which provides further dimensions. Each chapter has been rigorously reviewed and revised to represent the current state of knowledge and to demonstrate thoughtful discussion among researchers, healthcare professionals, policymakers, and technology developers.

We acknowledge the contributions of all authors, reviewers, and technical advisors who shared their expertise and time to enrich the quality of this volume. Special thanks are extended to Ms. Nina Miocevic, Publishing Process Manager, and to editorial assistants and collaborators for their unwavering support throughout the publication process.

We hope this book serves as a valuable reference and an inspiration for future research and innovation in the secure and effective use of Electronic Health Records within complex healthcare ecosystems.

Venkata Krishna Parimala
Department of Computer Science,
Sri Padmavati Mahila Visvavidyalayam,
Tirupati, India

Section 1

Cybersecurity and Risk Management in EHR Systems

Chapter 1

Assuring Patient Privacy and Safety through Robust Enterprise Cyber Risk Management (ECRM)

Sakshi Porwal and Bob Chaput

Abstract

As Electronic Health Records (EHRs) and advanced technologies such as quantum computing, artificial intelligence/machine learning, 5G, data analytics, cloud, blockchain, or IoT increasingly replace traditional IT applications, the ongoing digitization necessitates improved Enterprise Cyber Risk Management (ECRM) programs and the evolution of healthcare organizations' ECRM program and cybersecurity strategy. Cybersecurity protects investments and business value and plays a crucial role in patient safety. Robust, progressive ECRM programs and cybersecurity strategies are needed to save patients' lives and prevent harm resulting from compromised confidentiality, integrity, and availability of healthcare data, systems, and devices. This chapter will detail the steps organizations must take to develop and document their ECRM program and cybersecurity strategy, covering the entire data lifecycle from creation to storage to destruction. It will explain how ECRM must be integrated into business strategy and will provide insights on establishing, implementing, and maturing your cyber program and strategy.

Keywords: cybersecurity, privacy, enterprise cyber risk management, data management, data minimization, cybersecurity strategy, patient safety, healthcare data

1. Introduction

The shift toward Electronic Health Records (EHRs) and advanced technologies like quantum computing, artificial intelligence (AI), machine learning (ML), 5G, data analytics, cloud computing, blockchain, and the Internet of Things (IoT) has revolutionized healthcare. These technologies bring unprecedented opportunities for improving patient care, streamlining operations, and enhancing decision-making processes. For example, AI can now assist in diagnosing diseases earlier, while IoT devices can monitor patient vitals remotely in real-time, leading to better healthcare outcomes. However, the rapid integration of these technologies also introduces systemic risks. Recent trends, such as AI-powered phishing attacks and ransomware-as-a-service, highlight how cybercriminals are leveraging emerging technologies to exploit vulnerabilities in healthcare systems.

The primary concern with EHRs is the potential for compromises of confidentiality, integrity, or availability of patient data and the systems and devices that create, receive, maintain, or transmit this data. The growing complexity of healthcare data, systems, and devices and the increasing use of interconnected devices greatly expand the attack surface, making it easier for cybercriminals to find vulnerabilities. EHRs consolidate vast amounts of sensitive information, including personal identification details, medical histories, treatment plans, and financial data. Cybercriminals view this information as a valuable target, leading to increased data breaches and cyberattacks on healthcare organizations. When unauthorized access occurs, it can lead to identity theft, financial fraud, and the unauthorized sharing of sensitive health information, severely damaging patients' trust in their healthcare providers.

Additionally, the integrity of healthcare data is crucial. EHRs must ensure that patient information remains accurate and unaltered. Any data tampering or corruption, whether malicious or accidental, can lead to misdiagnoses, inappropriate treatments, and potentially life-threatening situations. Healthcare providers must implement strong integrity controls, including cryptographic signatures and real-time auditing, to prevent data manipulation. For example, if a patient's allergy information is altered or deleted, the administration of contraindicated medications could result in severe allergic reactions or even fatalities.

The availability of EHR systems is equally important. Healthcare providers rely on the constant availability of these systems to access patient information and deliver timely care. Ransomware attacks, which encrypt data and demand payment for its release, have increasingly targeted healthcare organizations. Such attacks can disrupt access to critical patient information, delaying treatments and compromising patient safety. Moreover, prolonged downtime in EHR systems can overwhelm healthcare staff, forcing them to revert to outdated paper records, which are prone to human error and can further endanger patient safety.

In 2014, in the U.S., the Federal Bureau of Investigation (FBI) issued a Private Industry Notice (PIN) warning of increased cyber intrusions against healthcare systems [1]. The alert cited several reasons for the increase, including the transition from paper records to EHRs, a higher financial payout for medical records on the black market, and "lax cybersecurity standards" in the industry [1]. The alert went on to state that "the health care industry is not as resilient to cyber intrusions" as other critical infrastructure sectors (e.g., energy, financial services, transportation systems, etc.) and that healthcare is "poorly protected and ill-equipped to handle new cyber threats exposing patient medical records." Similarly, in the EU, healthcare data breaches have been rising rapidly. According to the European Union Agency for Cybersecurity (ENISA), in 2022, healthcare was the fifth most affected sector by cyberattacks in Europe, with attacks increasing by 60% compared to the previous year. One of the most notable breaches occurred in Ireland in 2021, when the Health Service Executive (HSE) was hit by a ransomware attack, forcing the shutdown of national IT systems and disrupting healthcare services across the country. The attack affected around 700GB of data, including sensitive patient records [2]. The health sector desperately needs a cyber hygiene injection as cyber attacks continue to increase [3]. In Germany, a 2022 report revealed that nearly 40% of hospitals had experienced a cyberattack in the past year, further exposing vulnerabilities in the healthcare system [4]. In the United States, 144.4 million records were reported breached in 2023, an increase of 156% vs. 56.5 million in 2022 [5]. Similarly, in the EU, the average cost of a healthcare data breach was estimated at €4.27 million in 2023, highlighting the financial burden associated with weak cybersecurity practices in the sector [6].

The volume, visibility, vulnerability, and value of healthcare care systems and devices continue to explode. Integrating emerging technologies like IoT devices and AI/ML systems adds complexity to the security landscape. IoT devices like connected medical devices and wearable health monitors can provide real-time data and improve patient outcomes. However, these devices can also serve as cyberattack entry points if not adequately secured. Twenty-five years ago, the phrase "Internet of Things (IoT)" had not yet been coined [7]. Compare that to today: 87% of healthcare organizations expect to implement Internet of Medical Things (IoMT) technology by the end of 2019 [8]. It has been estimated that more than 3.7 million IoMT devices are in use today [9]. These IoT-enabled devices, if compromised, not only expose patient data but can also directly impact patient safety, if an attacker were to alter the functioning of a connected pacemaker or insulin pump.

AI/ML systems, which analyze vast datasets to provide insights and predictions, are only as reliable as the data they process. If the data integrity is compromised, the outputs can be misleading, leading to incorrect clinical decisions and patient harm. For instance, AI-driven diagnostic tools that rely on tampered or incomplete data can lead to misdiagnoses, delayed treatments, or inappropriate medical interventions.

The unintended consequences of digitizing the healthcare industry include operational disruption, greater legal and compliance exposures, new patient safety concerns, patient trust issues, potential reputational damage, possible medical malpractice lawsuits, and supply chain or third-party exposure. As these consequences accumulate, healthcare providers are increasingly faced with the need for robust risk management strategies that encompass not only IT systems but also human and operational factors. With so much at stake, why is healthcare so far behind other industries in terms of cybersecurity?

Healthcare has lagged behind other industries in adopting digital technology in general. The banking industry, for instance, went digital decades before healthcare did. The first automatic teller machine (ATM) that allowed customers direct access to their accounts was unveiled by Lloyds Bank in the United Kingdom in the early 1970s [10]. That was 35 years before EHRs were starting to be implemented. The healthcare industry's lack of cybersecurity preparedness is commonly attributed to the following:

- *Lack of proactive cybersecurity frameworks:* Healthcare providers tend to implement cybersecurity measures reactively after breaches occur, rather than embedding security by design.

- *Lack of security talent:* Most health delivery organizations lack full-time, qualified cybersecurity personnel.

- *Legacy equipment:* Many healthcare organizations are using equipment that is running on old, unsupported, and vulnerable operating systems.

- *Premature connectivity:* In the healthcare industry's rush to go live, cybersecurity concerns were often placed on the back burner or treated as an afterthought. Connectivity was put in place before consideration of secure design and implementation.

- *An open, sharing culture:* To deliver effective care, data has to be easily accessible to clinicians. There is sometimes tension between the accessibility needed for timely care delivery and the privacy and security practices required to protect patient data.

- *Viewing cybersecurity as an IT challenge:* There is a tendency to approach cybersecurity reactively (rather than proactively) and as an IT responsibility (rather than an enterprise-wide responsibility), which is expected in the healthcare industry.

- *Lack of understanding of the risks cyber threats pose:* Many in the healthcare industry, from clinicians in the field to organizational leadership, lack awareness and complete understanding of cyber risk [11].

The combination of increasing cyber-attacks on healthcare organizations, healthcare's attack surface exploding, the healthcare industry's lack of preparedness, risk illiteracy, lack of C-suite and board engagement, and lack of appreciation of business value in strong ECRM programs and cybersecurity strategies create significant patient privacy and safety issues and other challenges related to the deployment and operation of EHRs.

2. Actions organizations must take to assure patient privacy and safety

Healthcare organizations must develop and implement robust Enterprise Cyber Risk Management (ECRM) programs and cybersecurity strategies to address these challenges and safeguard patient privacy and safety. These strategies must encompass the entire data lifecycle, from creation to storage to destruction, ensuring data is protected at every stage. They must also include all the systems and devices associated with this data. In the EU, the General Data Protection Regulation (GDPR) has set a high bar for how organizations handle personal data, imposing stringent requirements for the security of health data. Non-compliance can result in penalties of up to 4% of global annual turnover, making ECRM a crucial element for healthcare organizations [12] .

2.1 Challenges conducting robust ECRM

Before enumerating specific, tangible actions an organization must take, consider some of the challenges of conducting ECRM. First, cyber risk management is new to all industries, including healthcare. Risk management is certainly not a new concept in the healthcare industry. Healthcare organizations have been managing risks related to patient safety and professional liability for a long time. What's changed in healthcare enterprise risk management is that enterprise cyber risk management has moved to the forefront very quickly. Cyber risk wasn't even on the healthcare industry radar 20 years ago. With the digitization of healthcare, cyber risk has become a risk with the potential to impact every other type of business risk a healthcare organization manages.

Today, cyber risk management involves two dimensions in most countries: (1) regulatory compliance risk management (think: GDPR or HIPAA) and (2) security risk management. While the courts, regulators, and legislators continue to raise the bar regarding compliance, many healthcare organizations still lag in complying with the myriad of regulations they must abide by. For instance, a 2022 report found that

63% of healthcare organizations in the EU had suffered a data breach in the past year, yet many still struggle with GDPR compliance, particularly in terms of encryption and data minimization practices [13]. Security risk management requires that organizations understand risk, in general, and cyber risk, particularly when it comes to healthcare data, systems, and devices. Risk illiteracy pervades the industry, and few so-called cybersecurity experts can clearly explain the difference between a risk, a threat, and a vulnerability.

To make matters even more interesting, many healthcare professionals equate compliance to security. Simply ticking off boxes on a controls checklist or list of best practices and calling it good does not translate into adequate security. It is a misconception that compliance risk management and security risk management are synonymous. Noncompliance with international, federal, and state laws, rules and regulations can lead to fines, penalties, and corrective actions being levied against an organization. Compliance is only one part of a much bigger cyber risk management picture. A compliance risk can trigger a security risk and vice versa.

As an example of the connection between regulatory compliance and practical implementation of Enterprise Cyber Risk Management (ECRM), consider the case of US-based Anthem, Inc. The largest settlement amount negotiated by US Department of Health and Human Services Office for Civil Rights (OCR) to date was the result of the largest breach of healthcare data up to that time. On "February 13, 2015, HHS opened a compliance review based on February 4, 2015, media reports and information on Anthem's website that Anthem had experienced a 'sophisticated external cyber attack.' On March 13, 2015, HHS notified Anthem that cyber-attackers gained impermissible access to the ePHI of 78,800,000 individuals maintained on Anthem's enterprise data warehouse. On February 23, 2015, HHS notified Anthem of its investigation regarding Anthem's compliance with the HIPAA Rules."

The settlement agreement listed five potential violations of the HIPAA regulations:

- Failure to conduct an enterprise-wide risk analysis
- Failure to regularly review records of information system activity
- Failure to identify and respond to suspected or known security incidents
- Failure to implement sufficient technical policies and procedures to protect ePHI from impermissible access
- Failure to prevent unauthorized access[1]

Anthem paid the highest OCR HIPAA negotiated settlement amount to date: $16 million. In addition, Anthem incurred costs of $2.5 million to engage expert consultants and $31 million in initial notification costs to victims. Seven state insurance commissioners reported that Anthem was spending more than $260 million on overdue security-related measures. The failure to establish, implement, and mature a

[1] U.S. Dep't of Health and Human Servs., Press Office, Anthem pays OCR $16 Million in record HIPAA settlement following largest health data breach in history - October 15, 2018, *available at* https://www.hhs.gov/hipaa/for-professionals/compliance-enforcement/agreements/anthem/index.html.

robust ECRM program resulted in Anthem incurring unexpected, out-of-pocket costs exceeding 400 million dollars.

Another challenge organizations face in conducting ECRM is the recognition that cyber risk management is not an IT problem; it's an existential business risk for most healthcare organizations and a patient safety issue. Cyber risk does not exist in isolation. A successful cyber attack on your organization does not just threaten your organization's PHI or other sensitive data; cyber risk is inextricably linked to your organization's different risks. A successful cyber attack could trigger an enforcement action by a regulatory body. Ransomware attacks lead to significant business interruption and an expensive payoff to resume operations. In the UK, for example, the WannaCry ransomware attack in 2017 severely impacted the National Health Service (NHS), resulting in 19,000 canceled appointments, including critical surgeries, highlighting the potential for patient harm [14]. Other consequences might include damage to your organization's reputation, patient death, lost patients and revenues, a medical malpractice lawsuit, or a negligence lawsuit against C-suite executives and board members.

The last challenge discussed is that the "mysticism" surrounding cyber often discourages the critical engagement required of the healthcare organization's C-suite and board of directors. Too many members of the C-suite and board of directors believe they need to be technical experts to engage in ECRM discussions. You do not need to be a technical expert to provide effective ECRM leadership and oversight for your organization. Nothing about cyber risk makes it more complicated than the other complex risks that hospital and health system C-suite executives and board members already address. For instance, boards in the EU are increasingly expected to engage in cybersecurity oversight, as reflected in the Network and Information Systems (NIS) Directive, which imposes security requirements on critical sectors, including healthcare [15]. Remember, for the board, the mantra is "eyes open, nose in, and fingers out" regarding oversight responsibilities [16]. Whether it is the risk of a competitor stealing market share, the risk of the organization's credit rating being lowered, the risk of non-compliance with regulations, the risk of medical malpractice, or the risk of spills and falls in the hospital, risk is all about loss or harm. To begin, you need to understand what assets are at risk and the frequency and severity of risk events that may cause loss or damage to these assets. That's business language, not technical language.

Without appropriate board oversight and C-suite leadership, decisions about treating cyber risk will default to the wrong level of the organization. No disrespect is intended toward your Chief Information Security Officer (CISO), Chief Compliance Officer (CCO), or Chief Information Officer (CIO), but these individuals often do not have the full strategic context of your organization. An enterprise-wide, strategic perspective—held by the C-suite and board—is essential to implementing an effective ECRM program. A 2022 ENISA report highlighted that leadership involvement is key to reducing cyber risk in healthcare, with organizations that had board-level cybersecurity engagement experiencing fewer incidents and faster recovery times [17].

2.2 Actions

Successful organizations treat their work to establish, implement, and mature their ECRM program as transformational and focus on five core capabilities: governance, people, process, technology, and engagement. Every organization is at a different level of maturity regarding its privacy and ECRM programs. The following recommendations include five strategic and several tactical actions designed to guide at all maturity levels.

2.3 Recommended strategic actions

2.3.1 Establish appropriate governance

The first critical core capability centers on governance. Governance is a system of processes and controls that ensures that stakeholder needs, conditions, and options are evaluated to determine balanced, agreed-upon enterprise objectives, setting direction through prioritization and decision-making, and monitoring performance and compliance against agreed-upon direction and goals.

In the context of ECRM governance, the board plays a vital role in setting direction and providing ongoing oversight. It is the board's responsibility to ensure that the organization's cyber risk management aligns with its broader goals and that there is an understanding of how cybersecurity impacts patient safety and trust. This includes the adoption of a risk management framework that defines roles, responsibilities, and accountability across the organization. The board communicates, "This is where we are going (with ECRM), and this is why we are going there." C-suite executives such as the CISO and CCO are responsible for executing these strategic decisions. For example, in the UK, the NHS introduced the "Data Security and Protection Toolkit" as a comprehensive governance strategy to ensure healthcare organizations properly manage patient data [18]. This toolkit ensures compliance with regional laws while embedding data governance at every level.

Furthermore, governance also involves ensuring compliance with regional and global regulations like the GDPR in Europe or HIPAA in US, which emphasizes not only protecting personal health data but also ensuring organizations are accountable and transparent in their data-handling practices. Compliance requirements include conducting data protection impact assessments (DPIAs) and appointing Data Protection Officers (DPOs) to monitor compliance. Inadequate governance can lead to significant fines under these regulatory frameworks. In 2021, a French hospital was fined €400,000 for GDPR violations following a data breach that exposed sensitive patient information, highlighting the need for strong governance [19].

To further strengthen governance, healthcare boards must also stay informed about emerging cyber threats. This includes leveraging cybersecurity dashboards and reports provided by the CISO or CIO to make data-driven decisions about risk management priorities. For instance, by regularly reviewing cybersecurity performance metrics, boards can identify trends, potential vulnerabilities, and areas for improvement, allowing them to allocate resources more effectively.

C-suite executives, including the CISO and CCO, are responsible for executing the board's strategic decisions, driving the implementation of ECRM strategies, and ensuring organizational adherence to compliance mandates. These executives ensure that governance frameworks are operationalized, and that cyber risk management is ingrained in daily activities across the organization. A strong governance framework is also crucial in managing third-party risks, especially in the age of cloud computing and vendor-managed services, ensuring that external partners meet the organization's security and compliance standards.

In addition to formal governance, healthcare organizations should consider establishing cybersecurity or risk committees that regularly report to the board. These committees could consist of key stakeholders, including IT technical staff, legal, and clinical leaders, ensuring diverse perspectives on risk management. Technical staff are responsible for executing their authorized tasks under the governance structure established by the C-suite executives and the board. These tasks typically include but are not limited to risk assessments, security architecture, security engineering, and security operations.

2.3.2 Resource skilled people

The second critical core capability focuses on *people*. An ECRM program's success is heavily dependent on having a well-equipped workforce with the right mix of technical and managerial expertise. The individuals managing these programs must have the necessary skills, knowledge, experience, and passion to drive cyber risk management effectively. This encompasses professionals at all levels, from strategic leadership in the C-suite to frontline cybersecurity analysts and IT support staff who monitor systems daily. In many healthcare organizations, however, there is a significant shortage of qualified cybersecurity talent, which has led to major challenges in securing sensitive patient data. According to a 2021 $(ISC)^2$ report, the global cybersecurity workforce gap exceeded 2.7 million professionals, a challenge acutely felt in the healthcare sector [20].

Healthcare organizations face the difficulty of both attracting and retaining top cybersecurity talent. The highly competitive market for cybersecurity professionals, compounded by the complexity of healthcare systems and the critical nature of the data involved, makes it even more difficult for healthcare organizations to fill these roles. To address this shortage, many organizations have opted to outsource certain aspects of their ECRM programs, bringing in third-party experts or managed service providers (MSPs) to assist with cybersecurity monitoring, incident response, and compliance management. For example, some healthcare providers in Europe have begun collaborating with cybersecurity firms to ensure 24/7 monitoring of critical systems, especially after high-profile breaches, such as the 2020 ransomware attack on Finland's Vastaamo psychotherapy center, which exposed sensitive patient therapy notes [21].

Appropriately resourcing an ECRM program also involves building internal capabilities. Organizations must invest in ongoing training and education to develop a risk-aware workforce. This includes training non-technical staff, such as nurses, doctors, and administrative personnel, who are often the first targets of phishing or social engineering attacks. These groups must be educated on how to recognize suspicious emails, prevent malware infections, and follow data protection best practices. For example, in the United Kingdom, the NHS has mandated cybersecurity training for all staff members following the 2017 WannaCry ransomware attack, which severely disrupted healthcare services [22].

At a higher level, organizations need to create leadership roles that are dedicated to cybersecurity and risk management. Positions such as the CISO, CCO, and DPO are vital in driving cybersecurity strategy and ensuring regulatory compliance. These leaders are responsible for not only setting the strategic direction for ECRM but also ensuring that the right resources are allocated to manage risks effectively. However, many healthcare organizations still do not have a dedicated CISO, which can lead to fragmented approaches to risk management and missed opportunities to integrate cybersecurity with broader business goals.

Resourcing an ECRM program means putting the right people in place at every level, from leadership to the tactical frontline. This often involves a mix of internal and external resources. Leveraging internal resources means hiring or upskilling staff to take on cybersecurity roles, while also fostering a culture of security awareness across the organization. Creating a risk-aware culture is essential, as the success of any cybersecurity program depends on employees understanding and taking responsibility for their part in managing cyber risks.

At the same time, leveraging external resources can provide access to specialized skills and expertise that may not be available in-house. For example, smaller healthcare organizations or clinics that may not have the budget to hire a full-time cybersecurity team can rely on Managed Security Service Providers (MSSPs) to perform essential services such as vulnerability scanning, incident response, and compliance audits. Additionally, partnering with academic institutions to develop talent pipelines or offering internships for cybersecurity students are ways to address the talent gap in the long run. This approach has been particularly successful in the Netherlands, where healthcare organizations have developed partnerships with local universities to ensure a steady flow of trained cybersecurity professionals [23].

Finally, part of the leadership responsibility of C-suite executives and board members is to understand the value and benefits of the ECRM program to justify the necessary resources. This includes justifying budget allocations for hiring cybersecurity personnel, investing in continuous training, and supporting certification programs for existing staff. Leaders must advocate for ECRM as a strategic investment in the organization's resilience, particularly given the growing sophistication of cyber threats and the increasing regulatory burden. Failure to properly resource the ECRM program can leave healthcare organizations vulnerable, not only to financial loss but also to legal repercussions and reputational damage, as seen in cases where data breaches result in lawsuits or fines from regulatory bodies like the U.S. Department of Health and Human Services or the European Data Protection Board (EDPB).

2.3.3 Adopt industry-standard processes

The third critical core capability is process. Processes are essential for ensuring consistency and accountability in cybersecurity efforts. At the most basic level, a process is a specific way of doing something - be it monitoring systems, responding to incidents, or assessing risks. For organizations to effectively manage cyber risk, they need well-established processes that are clear, repeatable, and adaptable to the dynamic nature of the healthcare landscape. A common business definition of process from Wikipedia, is "[A] business process, business method, or business function is a collection of related, structured activities or tasks performed by people or equipment in which a specific sequence produces a service or product (that serves a particular business goal) for a particular customer or customers [24]."

In the context of ECRM, processes are critical for ensuring that cybersecurity measures are applied consistently across the organization. Healthcare organizations must adopt industry-standard processes, especially as the attack surface continues to grow with the integration of EHRs, connected medical devices (IoT), cloud platforms, and third-party vendors. Each of these introduces its own set of vulnerabilities, making it essential for organizations to adopt processes that address these risks comprehensively.

Organizations with a mature ECRM process have formal, well-documented, and consistently followed policies, procedures, and practices around risk management. These policies must be aligned with the organization's risk tolerance and regulatory obligations. For example, healthcare providers in the EU must ensure their risk management processes comply with the GDPR, which places strict requirements on how personal health data is processed, stored, and protected. Additionally, processes must be scalable to ensure that as the organization grows, these cybersecurity practices continue to protect sensitive data effectively.

A mature ECRM process ensures a risk management approach that is predictable, measurable, and controlled, aligning with continuous process improvement (CPI) principles. This means processes are not static—they must be regularly reviewed and updated based on new threats, technologies, and lessons learned from past incidents.

As with other core capabilities, healthcare organizations can benefit by referencing standards-based guidance on cyber risk management processes rather than creating their own from scratch. Established frameworks such as the NIST Cybersecurity Framework (CSF), ISO/IEC 27001, and CIS Critical Security Controls provide comprehensive guidelines for healthcare organizations to follow, ensuring they meet best practices in protecting sensitive health data.

The NIST CSF, for example, is widely adopted by U.S. healthcare providers and offers a set of standards and best practices for managing cybersecurity risk. ISO/IEC 27001, on the other hand, is an internationally recognized standard for information security management systems (ISMS) and provides healthcare organizations with a risk-based approach to managing their information security. Organizations that implement these standards can improve their cybersecurity posture, reduce the likelihood of breaches, and meet regulatory compliance requirements.

For healthcare organizations, conducting comprehensive risk assessments is critical to identifying potential vulnerabilities in their EHR systems and other digital technologies. Risk assessments must consider the full spectrum of threats, from insider threats and human error to sophisticated external cyberattacks like ransomware or distributed denial of service (DDoS) attacks. This includes evaluating the security posture of connected devices, cloud infrastructure, and third-party vendors. With more healthcare providers adopting cloud-based solutions for data storage and patient management, risk assessments must evaluate how these systems are secured and how data is encrypted both in transit and at rest.

Connected devices such as IoT-based medical devices also present unique security challenges. Vulnerabilities in these devices can be exploited to gain access to hospital networks, disrupt services, or tamper with critical patient care. A formal process must be in place for regularly updating and patching these devices to prevent such risks. For example, after vulnerabilities in certain pacemakers were discovered, hospitals implemented stricter security protocols to ensure devices were updated regularly to prevent unauthorized access [25].

Once risks are identified, they must be prioritized based on their potential impact on patient safety and privacy. Healthcare organizations should establish a formal risk management process that categorizes risks by severity and potential harm. This allows leaders to allocate resources and focus on addressing the most critical risks first, such as those that could lead to data breaches, compromised patient safety, or regulatory violations.

Cybercriminals constantly adapt their tactics, and new technologies often introduce unforeseen vulnerabilities. Therefore, regular assessments ensure that the organization's risk management processes are keeping pace with the changing threat landscape. The Office for Civil Rights (OCR)'s publication. "Guidance on Risk Analysis Requirements under the HIPAA Security Rule," provides specific directives for healthcare organizations on how to conduct these assessments [26]. These guidelines are supported by two NIST publications: NIST-SP 800-39

Managing Information Security Risk and NIST SP 800-30 Guide for Conducting Risk Assessments that offer a comprehensive approach to identifying, analyzing, and managing information security risks [27].

Risk assessments should also extend to third-party vendors, as supply chain risks are becoming increasingly prevalent in the healthcare sector. With many healthcare providers outsourcing data storage, cloud services, and even cybersecurity operations, a robust process must be in place for assessing the security practices of third-party vendors. The SolarWinds attack in 2020 underscored how vulnerabilities in third-party software can have far-reaching consequences for organizations, including those in the healthcare sector [28].

2.3.4 Employ relevant technology

The fourth critical core capability is technology. Technology serves as the backbone for any robust ECRM program, ensuring the efficient management and security of vast amounts of data and systems within healthcare organizations. Nearly all healthcare organizations already employ technology tools and automation to streamline clinical, administrative, and operational processes. Technological tools can also enable efficient ECRM workflows. However, the rapid evolution of cyber threats means that these organizations must continuously upgrade and adapt their technology stacks to stay ahead of emerging risks. Advanced technological solutions can help automate complex risk assessments, monitor system vulnerabilities, and predict potential breaches before they happen.

More importantly, technology tools are essential for the scaling ECRM to meet the growing demands of healthcare industry. Even the smallest healthcare organization has thousands, if not tens of thousands, of information assets (data, systems, and devices). For example, with the increased use of connected medical devices (IoT), telemedicine platforms, and cloud-based data storage, the digital footprint of healthcare providers is expanding exponentially [29]. Multiply that by the other variables comprising the risk scenario {asset-threat-vulnerability}, and a typical healthcare organization may look at more than 100,000 different risk scenarios that must be analyzed to complete a comprehensive, enterprise-wide risk assessment.

It is simply impossible to complete and maintain a comprehensive, enterprise-wide risk assessment without an appropriate technology and automation solution. This includes tools for vulnerability management, data protection, continuous monitoring, and automated reporting systems. Since the results of your risk assessment serve as the foundation for your ECRM strategy, it is critical to have the right ECRM technology solution in place. That means using standards-based technology. The advantage of using standards-based technology is that standards (such as the NIST Cybersecurity Framework or ISO/IEC 27001) have been developed, vetted, and successfully deployed across multiple organizations in multiple industries worldwide. These standards provide a structured approach to identifying, assessing, and managing risks and are internationally recognized, ensuring that organizations meet both regulatory and best practice requirements. Standards-based technology delivers consistent, predictable, repeatable, and measurable results, with the added benefit of explicit recognition by regulatory agencies.

One of the key components of a technology-driven ECRM program is encryption. Implementing robust encryption protocols is essential to protect data in transit and at rest. Encryption helps ensure that even if sensitive data is intercepted, it cannot be read or used by unauthorized individuals. Only authorized personnel should have access to sensitive patient information. Healthcare organizations must adopt multi-factor authentication (MFA) and role-based access controls to ensure access is granted based on the principle of least privilege. This means that each user has only the access they need to perform their job, minimizing the potential damage from a compromised account. Logging and monitoring access to EHRs can help detect and respond to unauthorized access attempts. Continuous monitoring solutions, including Security Information and Event Management (SIEM) systems, can alert security teams in real time if suspicious activity is detected.

In addition to traditional security technologies, advanced technologies such as AI and ML are transforming cybersecurity efforts. AI-powered threat detection systems can identify anomalies and potential threats in real time, enabling faster response times. For instance, AI can detect abnormal behavior patterns—such as unauthorized access to medical records outside normal working hours—signaling a potential security breach. ML algorithms can learn from past incidents to predict and prevent future attacks. For example, healthcare systems are increasingly using AI to detect ransomware before it can fully infiltrate a network, significantly reducing the impact of such attacks.

Blockchain technology can provide secure and immutable records of data transactions, ensuring data integrity and traceability. By using blockchain, healthcare organizations can create a tamper-proof chain of custody for medical data, ensuring that all changes are tracked and verified. This is especially useful for sharing data across different healthcare providers or with insurers, where maintaining data integrity is crucial.

Quantum computing, while still in its infancy, holds the promise of advanced encryption methods that can withstand future cyber threats. Quantum encryption methods, which leverage the principles of quantum mechanics, could potentially make data breaches nearly impossible by detecting any attempt to intercept or alter data in real time. While we are still years away from widespread use, organizations should start exploring how quantum technologies could be integrated into their long-term cybersecurity strategies.

Ultimately, the technology and automation tools you use to support your ECRM program will range from strategic-level solutions (such as a risk assessment solution) to operational-level solutions (such as a SIEM system). Risk assessment solutions, in particular, provide a framework for continuously identifying, evaluating, and mitigating risks across the healthcare environment, integrating with other tools to create a holistic view of organizational security. SIEM systems help aggregate and analyze security data from multiple sources, enabling proactive threat detection and rapid response.

The technology with the most significant relevance to the C-suite and board is the risk assessment/management solution. This solution provides the high-level insights necessary for executives to make informed decisions regarding cybersecurity investments, incident response strategies, and regulatory compliance efforts.

The ECRMS solution provides the foundation for the ECRM program. It should include appropriate C-suite and board-level dashboards and reports that give the C-suite and board the information you need to execute your ECRM leadership and oversight responsibilities. These dashboards should display real-time risk metrics, compliance status, and incident response times, providing a clear picture of the organization's security posture.

2.3.5 Ensure organizational engagement

The fifth and final core capability is engagement. For an ECRM program to be truly effective, engagement must go beyond the IT or cybersecurity team—it needs to be embedded across the entire organization. The success of your ECRM program depends on the extent to which the entire organization is actively engaged in its principles and practices. Cyber risk management requires a holistic, organization-wide approach where every department, from clinical staff to administrative functions, understands and participates in maintaining a secure environment. ECRM is not "just an IT problem." Nor is it "just a compliance problem," "just a privacy problem," or "just a chief risk officer problem." Cyber risk management is an existential risk management issue with consequences that can impact your patients and every stakeholder in your organization.

Engagement begins with clear communication from the top. Leadership—especially the C-suite and board—must communicate that cybersecurity is a priority for the entire organization. It should be seen not only as a protective measure but as a driver of trust, quality care, and operational resilience. Without a strong tone from the top, engagement will be weak, and cybersecurity efforts may be seen as just a technical function rather than a key organizational responsibility. This top-down support encourages every employee to understand their role in the cybersecurity ecosystem.

It follows that everyone in your organization has a role to play in your ECRM program. This involves integrating cybersecurity responsibilities into day-to-day tasks across all departments. For example, frontline healthcare workers should be trained to recognize phishing attacks, report suspicious activity, and handle patient data securely. Administrative staff should be aware of secure document handling procedures, and IT teams should ensure system configurations align with cybersecurity best practices. Engaging all employees also requires building a culture of awareness where cyber risk is discussed regularly, and employees feel empowered to contribute to the organization's defense. Healthcare providers in Denmark, for example, have implemented mandatory cybersecurity training programs, where every employee must complete annual training on data protection and information security [30].

Even if your C-suite and board are providing appropriate leadership and oversight, if your organization's other executives, managers, and workforce members are not engaged in your ECRM program, it will fail. Engagement must be reinforced at every level of management. Department heads and line managers should take ownership of cyber risk as part of their overall risk management responsibilities. This includes identifying risks specific to their department (such as insecure patient data access in clinical units or unencrypted communications in administrative departments) and reporting them through proper channels. Without engagement and ownership of risks by line of business, process, and functional leaders, people will make risk-related decisions without the whole strategic business view. By encouraging ownership of cyber risk at every level, organizations can ensure that cybersecurity is not siloed within the IT department but becomes an integral part of the organization's risk culture.

Another key aspect of organizational engagement is cross-functional collaboration. Effective cyber risk management often involves collaboration between different teams—legal, compliance, HR, IT, and clinical operations, to name a few. Establishing cross-functional risk committees or working groups allows various parts of the organization to share insights, discuss vulnerabilities, and develop integrated strategies for addressing cyber threats. These groups can also help in conducting tabletop exercises to simulate cyber incidents, allowing employees across functions to practice their response roles. For instance, a recent report from the UK NHS emphasized the importance of collaborative cybersecurity drills involving both IT teams and clinical staff to ensure a coordinated response to potential breaches [31].

All organizations have (or should have) an enterprise-wide risk management plan that describes the broad, strategic objectives to be pursued. This plan should explicitly integrate cyber risk as a core component of overall risk management. It's not enough to treat cybersecurity as an add-on or an isolated initiative—it should be embedded within the organization's broader risk management framework. At the same time, requiring departments to develop their own ECRM plans (within the context of the organization's overall ECRM framework and strategy) can help enforce accountability for risk management throughout the organization. By empowering each department to create tailored risk management plans, employees are encouraged to take an active role in safeguarding their specific environments. For example, the finance department's plan may focus on securing payment processing systems, while clinical departments may focus on protecting patient records and IoT-connected medical devices.

Measuring engagement is also crucial. Organizations should regularly assess how well employees are engaging with cybersecurity initiatives, through both qualitative and quantitative measures. This can include employee surveys on cybersecurity awareness, tracking completion rates for mandatory cybersecurity training, and analyzing the frequency of reported security incidents or near-misses. Regular engagement assessments allow leadership to identify areas where additional training or focus is needed and to continuously improve the ECRM program.

2.4 Recommended tactical actions

2.4.1 Plan your incident response

A well-defined incident response plan is critical to minimizing the impact of cyber incidents. Incident response planning is not just about reacting to an attack but being fully prepared to act swiftly and limit potential damage. Healthcare organizations should establish a dedicated incident response team trained to handle data breaches, ransomware attacks, and other cybersecurity incidents. The incident response team should include not only IT and security professionals but also representatives from legal, compliance, public relations, and clinical operations to ensure a holistic response. This multi-disciplinary approach ensures that all aspects of an attack, from patient safety to regulatory reporting, are addressed.

The plan should include clear, actionable steps for containment, investigation, communication, and recovery. For example, if a ransomware attack occurs, the containment strategy may involve isolating affected systems from the network to prevent the spread of the attack. Investigation procedures should detail how to identify the scope and source of the breach, while communication protocols should outline how and when to inform patients, staff, and regulatory bodies of the incident. Communication is particularly critical in healthcare, where transparency and patient trust are paramount.

In a 2020 report on the healthcare sector, the FBI recommended that healthcare organizations include proactive media strategies in their incident response plans, allowing them to address public concerns quickly and avoid reputational damage [32].

Regular drills and simulations can help ensure that the response team is prepared to act swiftly and effectively in the event of an attack. Tabletop exercises should simulate various types of attacks, from ransomware to insider threats, to test the team's ability to respond. Additionally, healthcare organizations can partner with external cybersecurity firms to conduct red team/blue team exercises, where attackers (the red team) simulate real-world attack scenarios while defenders (the blue team) work to stop them. These simulations help identify gaps in the incident response plan and improve the organization's overall readiness. For example, after a simulation exercise, a healthcare provider may find that they need to streamline communication channels between IT and clinical teams to prevent delays in notifying staff of a potential breach.

2.4.2 Integrate ECRM into business strategy

Cybersecurity must be considered a strategic priority and integrated into the overall business strategy. In the healthcare industry, where patient safety and data integrity are directly tied to operational success, integrating ECRM (Enterprise Cyber Risk Management) into business strategy is essential for long-term resilience. This involves aligning cybersecurity goals with organizational objectives and securing executive support. Healthcare providers should ensure that cybersecurity initiatives align with broader goals such as patient care excellence, regulatory compliance, and cost-efficiency.

By embedding cybersecurity into the organization's culture, healthcare providers can foster a proactive approach to risk management, encouraging employees to prioritize patient privacy and safety. For example, by tying cybersecurity efforts to patient care quality metrics, healthcare organizations can demonstrate that protecting patient data is not just a compliance issue but a key component of high-quality care. This helps build internal support for cybersecurity investments.

Securing executive support is critical. C-suite executives and board members must understand that cybersecurity is not a technical issue to be delegated solely to IT—it is a business issue that can affect everything from patient outcomes to financial performance. Executives should be involved in regular reviews of cybersecurity metrics, such as the number of incidents reported, time to respond, and compliance with regulatory requirements. For example, some hospitals in the U.S. have started integrating cybersecurity performance into their executive dashboards, ensuring that leadership has real-time visibility into potential risks [33]. This visibility helps leadership make informed decisions about resource allocation, technology investments, and risk mitigation strategies.

2.4.3 Train and educate your workforce

Employees play a crucial role in maintaining the security of EHR systems and other sensitive healthcare data. In fact, studies show that human error remains one of the leading causes of data breaches, underscoring the importance of a well-trained workforce. Regular training sessions should be conducted to educate staff on recognizing phishing attempts, reporting suspicious activities, and adhering to best practices for data handling. Given the complexity of the healthcare environment, it's essential to go beyond generic training programs. Training should be tailored to specific roles within the organization, focusing on the unique risks and responsibilities associated with each job function.

For example, frontline healthcare workers may need training on how to securely handle patient data when using mobile devices or telemedicine platforms, while administrative staff should be trained on secure email practices and how to avoid falling victim to social engineering attacks. Clinical staff may also need training on the secure use of IoT-enabled medical devices, which are increasingly becoming targets for cybercriminals. In 2020, the FDA released guidelines on medical device cybersecurity, urging healthcare providers to implement security measures and educate users on potential vulnerabilities [34].

Training should also focus on creating a culture of security awareness. Organizations can conduct phishing simulations to test employees' ability to recognize and respond to phishing attempts. These simulations not only reinforce training but also provide valuable insights into areas where additional education may be needed. For example, if a simulation reveals that many employees are clicking on fake phishing links, additional training on how to identify suspicious emails may be required. Ongoing education programs should be paired with regular updates on new cybersecurity threats, ensuring that staff remains vigilant in a constantly evolving threat landscape.

Furthermore, training should be complemented by clear reporting channels for employees to report potential cybersecurity incidents without fear of repercussions. Healthcare organizations that encourage open communication about cybersecurity risks are more likely to detect and mitigate threats early.

3. Benefits and value organizations can expect from a robust ECRM program

By implementing a robust ECRM program and cybersecurity strategy, healthcare organizations can expect to realize numerous benefits that directly impact patient privacy, safety, and overall organizational performance.

1. *Enhanced patient trust:* Patients who know their sensitive information is protected are more likely to trust their healthcare providers. This trust is essential for building long-term patient relationships and encouraging open communication. When patients feel secure about their data, they are more likely to share complete and accurate information, leading to better diagnoses and treatment plans. Furthermore, healthcare organizations that prioritize data privacy are better positioned to maintain their reputations, even in the event of a minor breach. By safeguarding patient data, organizations demonstrate their commitment to patient privacy, fostering confidence in their ability to provide safe and effective care.

2. *Improved patient safety*: A secure EHR system ensures healthcare providers access accurate and unaltered patient information. This reduces the risk of medical errors, misdiagnoses, and inappropriate treatments, ultimately enhancing patient safety. For example, secure EHRs prevent unauthorized modifications to critical data like allergy information or medication history, which could lead to life-threatening consequences. In high-stakes medical environments such as surgeries or emergency rooms, reliable access to accurate patient data can significantly improve care outcomes and reduce the potential for malpractice. Timely access to reliable data is crucial for delivering high-quality care, especially in emergencies where every second counts.

3. *Regulatory compliance*: Healthcare organizations in the United States are subject to strict regulations regarding patient data protection, such as the HIPAA. Organizations can ensure compliance with these regulations by implementing comprehensive cybersecurity measures, avoiding costly fines and legal repercussions. Failure to comply with data protection laws not only exposes organizations to financial penalties but also reputational damage that can deter patients from seeking care. A robust ECRM program also simplifies the process of conducting audits and reporting incidents to regulatory bodies, ensuring a more streamlined compliance process.

4. *Operational resilience:* A robust cybersecurity strategy enhances the resilience of healthcare organizations against cyber threats. By having a well-prepared incident response plan and strong security measures in place, organizations can quickly recover from cyber incidents and minimize disruptions to their operations. This operational resilience is crucial in an industry where delays or system outages can directly impact patient care, leading to serious health risks. Additionally, resilient systems can help healthcare organizations avoid extended downtime, maintaining their reputation as reliable care providers and ensuring patient trust remains intact. This resilience is crucial for maintaining the continuity of patient care and minimizing financial losses.

5. *Cost savings:* While investing in cybersecurity may seem costly, the long-term savings outweigh the initial expenses. Preventing data breaches and cyberattacks can save organizations from the financial losses associated with data recovery, legal actions, reputational damage, and regulatory penalties. Cyberattacks in the healthcare sector can lead to multimillion-dollar recovery efforts, especially if sensitive patient data is compromised or operations are severely disrupted. By proactively addressing potential vulnerabilities and implementing robust security measures, healthcare providers can avoid the steep costs of remediation, lawsuits, and regulatory fines. A proactive approach to cybersecurity is a cost-effective strategy for protecting the organization's assets and reputation.

6. *Innovation and growth:* By prioritizing cybersecurity, healthcare organizations create a secure foundation for adopting new technologies and innovations. With the confidence that patient data is protected, organizations can explore opportunities to integrate AI, IoT, and other advanced technologies into their operations. For instance, secure integration of IoT devices allows for real-time monitoring of patient vitals, improving diagnostic accuracy and patient outcomes. Moreover, AI-powered systems can enhance decision-making, automate routine tasks, and streamline operations when supported by a strong cybersecurity framework. This can lead to improved patient outcomes, enhanced operational efficiency, and a competitive edge in the healthcare industry.

4. Conclusion

In conclusion, the ongoing digitization of healthcare presents both opportunities and challenges. The increasing reliance on digital systems, from EHRs to connected medical devices, has created new vulnerabilities that cybercriminals are eager to exploit. However, it also offers healthcare providers a chance to improve care delivery, streamline operations, and engage with patients in more meaningful ways—provided they manage the associated risks effectively.

To successfully navigate this digital transformation, healthcare organizations must take proactive steps to mitigate risks, ensure patient safety, and build operational resilience. The following actionable roadmap outlines the critical priorities for organizations:

- *Establish robust governance:* Engage leadership at the C-suite and board levels to align cyber risk management with strategic goals, implement frameworks like NIST CSF, and regularly review threat intelligence to guide decision-making.

- *Invest in talent and training:* Build internal cybersecurity capabilities by hiring skilled professionals and conducting role-specific training for all staff while collaborating with external partners to address skill gaps.

- *Adopt comprehensive risk management processes:* Conduct regular risk assessments to identify vulnerabilities, prioritize risks using tools like heat maps, and establish a formal incident response plan to mitigate and recover from threats effectively.

- *Leverage advanced technology:* Use AI-driven tools for real-time threat detection, implement encryption and access controls for safeguarding sensitive data, and explore emerging solutions like blockchain and quantum encryption.

- *Foster organization-wide engagement:* Promote a culture of cybersecurity awareness through regular training, cross-functional simulations, and clear communication, ensuring every employee contributes to the organization's defense.

- *Integrate cybersecurity into business strategy:* Tie cybersecurity goals to patient care and operational objectives, allocate adequate budgets, and incorporate key metrics into executive dashboards for informed decision-making.

By following this roadmap, healthcare organizations can transform cybersecurity from a reactive necessity to a proactive enabler of trust, innovation, and operational resilience. These efforts are not just about safeguarding data; they are about saving lives and maintaining patients' trust in their healthcare providers. Beyond regulatory compliance, robust Enterprise Cyber Risk Management (ECRM) programs and strategies ensure that healthcare providers are prepared to face growing cyber threats while embracing digital innovation with confidence.

Ultimately, the healthcare organizations that prioritize cybersecurity as a core component of their digital transformation efforts will be better positioned to thrive in an increasingly interconnected and data-driven world, achieving their mission to deliver safe, reliable, and patient-centered care.

Author details

Sakshi Porwal[1*] and Bob Chaput[2,3]

1 Compunnel, Inc., Princeton, NJ, USA

2 Clearwater Security, Nashville, TN, USA

3 Center for Professional Education, The University of Texas at Austin, Austin, TX, USA

*Address all correspondence to: porwalsakshi@gmail.com

References

[1] Health Care Systems and Medical Devices at Risk for Increased Cyber Intrusions for Financial Gain. Private Industry Notification (PIN) #: 140408-009., Vols. Available from: https://info.publicintelligence.net/FBI-HealthCareCyberIntrusions.pdf, p. FBI Cyber Division. 2014

[2] Theocharidou M, Lella I. Threat Landscape: Healthcare Sector, European Union Agency for Cybersecurity (ENISA). ISBN 978-92-9204-638-5. Available from: https://www.enisa.europa.eu/publications/health-threat-landscape. 2023

[3] Scott J. GoodReads, Healthcare Cybersecurity Quotes. 2015. Available from: https://www.goodreads.com/quotes/tag/healthcare-cybersecurity [Accessed: August 31, 2019]

[4] Alder S. Global healthcare cyberattacks increased by 74% in 2022. The HIPAA Journal. 2023. Available from: https://www.hipaajournal.com/global-healthcare-cyberattacks-increased-by-74-in-2022/

[5] Portal B. Office for Civil Rights (OCR). Available from: https://ocrportal.hhs.gov/ocr/breach/breach_report.jsf, U.S. Department of Health and Human Services

[6] Greenlee M. Cost of a data breach 2023: Healthcare industry impacts. Security Intelligence. 2023. Available from: https://securityintelligence.com/articles/cost-of-a-data-breach-2023-healthcare-industry-impacts/

[7] Foote KD. A brief history of the internet of things. Dataversity. 2016

[8] Cohen JK. How is IoT used in healthcare: 7 statistics. Becker's Health IT & CIO Report. 2017. Available from: https://www.dataversity.net/brief-history-internet-things/

[9] Marr B. Why the internet of medical things (IoMT) will start to transform healthcare in 2018. Forbes. 2018. Available from: https://www.forbes.com/sites/bernardmarr/2018/01/25/why-the-internet-of-medical-things-iomt-will-start-to-transform-healthcare-in-2018/#3e9125284a3c

[10] Bátiz-Lazo B. A brief history of the ATM. The Atlantic. 2015. Available from: https://www.theatlantic.com/technology/archive/2015/03/a-brief-history-of-the-atm/388547/ Accessed: September 1, 2019

[11] Industry, report on improving Cybersecurity in the health care. Health Care Industry Cybersecurity Task Force. 2017. Available from: https://www.phe.gov/Preparedness/planning/CyberTF/Documents/report2017.pdf [Accessed: 31 August, 2019]

[12] GDPR Fines/Penalties. Germany: Intersoft Consulting; Available from: https://gdpr-info.eu/issues/fines-penalties/

[13] Alder S. 66% of healthcare organizations say patient care was disrupted by a cyberattack. The HIPAA Journal. 2023. Available from: https://www.hipaajournal.com/66pc-healthcare-organizations-patient-care-disruption-cyberattack/

[14] Executive NH. WannaCry cyber-attack cost the NHS £92m after 19,000 appointments were cancelled. National Health Executive. 2018 [Online]. Available: https://www.nationalhealthexecutive.com/articles/

wannacry-cyber-attack-cost-nhs-ps92m-after-19000-appointments-were-cancelled

[15] L. V. and. N. S. Tim Van Canneyt. The NIS 2 Directive – Implications for the Healthcare Sector and Manufacturers of Medical Devices. Fieldfisher; 2023. Available from: https://www.fieldfisher.com/en/insights/the-nis-2-directive-implications-for-the-healthcare-sector-and-manufacturers-of-medical-devices

[16] Mastersmith BABAL. Eyes Open, Nose in, Fingers out: Understanding and Managing the Potential Risks of Signing on as a Corporate Director. London, United Kingdom: Law Business Research; 2017. Available from: https://www.lexology.com/library/detail.aspx?g=f670c6ea-23cf-4c96-bd5b-e6d1b75bfde9

[17] ENISA Threat Landscape. 2022. ISBN: 978-92-9204-588-3. Available from: https://www.enisa.europa.eu/publications/enisa-threat-landscape-2022

[18] Data Security and Protection Toolkit. UK: NHS England; 2023. Available from: https://digital.nhs.uk/data-and-information/looking-after-information/data-security-and-information-governance/data-security-and-protection-toolkit

[19] Kiser M. Google Vs. GDPR: The Ripple Effect of the Biggest Data Protection Fine to Date. USA: Tech Radar; 2019. Available from: https://www.techradar.com/news/google-vs-gdpr-the-ripple-effect-of-the-biggest-data-protection-fine-to-date

[20] Cybersecurity Workforce Study. Looking deeper into the workforce gap. ISC2. 2023. Available from: https://www.isc2.org/Insights/2023/11/ISC2-Cybersecurity-Workforce-Study-Looking-Deeper-into-the-Workforce-Gap

[21] Corfield G. Hackers rummaged about in Finnish psychotherapy clinic – Now patients extorted with public data dump threats. The Register. 2020. Available from: https://www.theregister.com/2020/10/26/finland_psychotherapy_clinic_ransom_attack/

[22] Investigation: WannaCry Cyber Attack and the NHS. UK: National Audit Office; 2017. Available from: https://www.nao.org.uk/reports/investigation-wannacry-cyber-attack-and-the-nhs/

[23] H. Foundation. Collaboration to Improve Digital Security of the Life Sciences & Health Sector. Netherlands: HSD - Security Delta; 2023. Available from: https://securitydelta.nl/news/interviews/collaboration-to-improve-digital-security-of-the-life-sciences-health-sector

[24] Business Process. Wikipedia. Available from: https://en.wikipedia.org/wiki/Business_process

[25] Arghire I. St. Jude Medical recalls 465,000 pacemakers over security vulnerabilities. Security Week. 2017. Available from: https://www.securityweek.com/st-jude-medical-recalls-465000-pacemakers-over-security-vulnerabilities/

[26] Guidance on Risk Analysis. Washington, DC: US Department of Health and Human Services. 2019. Available from: https://www.hhs.gov/hipaa/for-professionals/security/guidance/guidance-risk-analysis/index.html

[27] Marron J. NIST's Planned Updates to Implementing the HIPAA Security Rule: A Cybersecurity Resource Guide. Washington, DC: NIST; 2023. Available from: https://www.nist.gov/blogs/cybersecurity-insights/nists-planned-updates-implementing-hipaa-security-rule-cybersecurity

[28] Strategic Threat Intelligence: Preparing for the Next "Solarwinds" Event. Washington, DC: American Hospital Association; 2021. Available from: https://www.aha.org/system/files/media/file/2021/04/hisac-aha-white-paper-strategic-threat-intelligence-preparing-for-next-solarwinds-event_0.pdf

[29] Kumar AKSVPBDGS-HKAASMSHM. Healthcare internet of things (H-IoT): Current trends, future prospects, applications, challenges, and security issues. Electronics. 2023;**12**(9):2050. DOI: 10.3390/electronics12092050

[30] Cyber and information security. Sundhedsdata Styrelsen. 2023. Available from: https://Security Analysis Centre for the health sector - The Danish Health Data Authority

[31] Lovell T. NHS England to adopt new cyber security framework. DIGITAL HEALTH. 2024. Available from: https://www.digitalhealth.net/2024/09/nhs-england-to-adopt-new-cyber-security-framework/

[32] Dalton C. Flash report: FBI warns of Ransomware assault on the healthcare and public health sector. PRO. 2020. Available from: https://tcblog.protiviti.com/2020/10/30/flash-report-fbi-warns-of-ransomware-assault-on-the-healthcare-and-public-health-sector/

[33] Boehm PMTPRRATSJ. Cyber Risk Measurement and the Holistic Cybersecurity Approach. Germany: McKinsey and Company; 2018

[34] FDA Outlines Cybersecurity Recommendations for Medical Device Manufacturers. US Food and Drug Administration; 2016. Available from: https://www.fda.gov/news-events/press-announcements/fda-outlines-cybersecurity-recommendations-medical-device-manufacturers

Chapter 2

Integrating Artificial Intelligence and Cybersecurity in Electronic Health Records: Addressing Challenges and Optimizing Healthcare Systems

Elena-Anca Paraschiv, Carmen Elena Cîrnu and Adrian Victor Vevera

Abstract

The digitalization of healthcare systems, particularly through Electronic Health Records (EHRs), presents both opportunities and challenges. This chapter delves into the transformative potential of integrating Artificial Intelligence (AI) with advanced cybersecurity measures in EHR systems. The impressive capabilities of AI models in data management, predictive analytics, and automation are explored for their role in enhancing patient outcomes and streamlining healthcare operations. The study addresses critical cybersecurity issues, including data breaches and ransomware, emphasizing the necessity of encryption, multi-factor authentication, and continuous monitoring. It examines how AI-driven threat detection and automated incident response can proactively safeguard sensitive patient data, also highlighting the challenges that may appear in the integration of AI into EHR systems along with addressing the need for robust interoperability standards and comprehensive data governance frameworks to mitigate cyber threats. The discussion extends toward the future vision that includes continuous innovation and strategic investment in AI and cybersecurity to create a more efficient, secure, and patient-centric healthcare environment. This comprehensive analysis highlights the synergistic potential of AI and cybersecurity in revolutionizing EHR systems and enhancing the overall quality of healthcare delivery.

Keywords: artificial intelligence (AI), cybersecurity, electronic health records (EHR), data integrity, interoperability

1. Introduction

The concept of only "electronic medical records" has become obsolete, giving way to the current trend of "electronic health records" or "patient records", which integrate clinical information from multiple sources, offering a complete view of the patient's health. Electronic health records (EHR) serve as the foundation for delivering effective and safe healthcare. As the medical industry continues to rapidly digitalize, the impact and the integration of EHRs are poised to become even more crucial and influential in the years ahead.

The healthcare industry is on the brink of a transformative revolution driven by the integration of artificial intelligence (AI) and cybersecurity into EHRs. As digital health data continues to proliferate, the need for sophisticated, secure, and efficient management systems becomes increasingly critical. EHR provides a comprehensive, real-time patient history that enhances clinical decision-making and patient outcomes. However, the integration of AI technologies introduces both profound benefits and significant challenges. AI has the potential to revolutionize healthcare by enabling predictive analytics, personalized treatment plans, and streamlined administrative processes. Yet, it also raises concerns regarding data security, patient privacy, and ethical considerations.

Simultaneously, the rise of cyber threats in the healthcare sector underscores the necessity of robust cybersecurity measures. Health data breaches can have devastating consequences, from financial losses to compromised patient trust. Therefore, integrating advanced cybersecurity practices is paramount to safeguarding sensitive health information. We aim to explore the dynamic intersection of AI and cybersecurity within EHRs, addressing the myriad challenges and uncovering opportunities for optimizing healthcare systems.

In this chapter, we will delve into the critical role of AI in enhancing EHRs, the cybersecurity challenges that must be addressed, and the strategies for integrating these technologies effectively. We will also examine regulatory and ethical considerations and the importance of user adoption. By understanding and addressing the challenges associated with AI and cybersecurity integration, healthcare organizations can unlock the full potential of EHRs, ultimately leading to improved patient care, operational efficiency, and a more secure healthcare environment.

2. AI and its potential in the management of EHRs

EHRs integrate a wide array of data essential for patient care, including medical history, demographics, medication records, physical measurements, laboratory results, vital signals, clinical notes, social history, and billing details. EHR systems facilitate the automatization of patient information, thereby improving clinical practice, supporting decision-making processes, and enhancing quality management and healthcare services [1]. The vast amount of data collected through EHR systems can play a significant role in improving patient care, optimizing clinical workflows, and contributing valuable insights for biomedical research and public health initiatives. This data [2] can be categorized into:

- *Structured data*: is stored in databases in predefined patterns, being quantifiable and easily retrievable. It is represented by key patient details, such as birth date, demographic details, prescribed treatments and medications, allergies, as well as weight, height, blood type, heart rate, or blood pressure. Structured data's standardized formats ensure consistency and accuracy across various healthcare systems, facilitating seamless data sharing and integration. It also supports advanced analytics, being easily integrated into AI-based algorithms in order to enable healthcare providers to identify patterns, predict patient outcomes, and make data-driven decisions efficiently.

- *Semi-structured data*: does not adhere to a strict format, being commonly utilized to enable data exchange and integration across various systems. This type of data gathers information based on disorder names, examination results, timestamps, and clinical guidelines and enables healthcare professionals to document complex medical information in a more organized manner than unstructured data, yet offers more flexibility than structured one.

- *Unstructured data*: is the most expansive and detailed form of data, encompassing a variety of narrative content such as medical notes, surgical documentation, radiological results, discharge reports, patient narratives, doctor's notes, and detailed descriptions of procedures. Additionally, there are also multimedia elements, such as medical images (X-rays, Computer Tomography images, etc.), audio recordings of consultations, or video recordings of different procedures. Although unstructured data is difficult to standardize and analyze, it offers in-depth insights into medical histories, clinical reasoning, and treatment results. This depth of information makes it invaluable for personalized patient care, in-depth clinical research, and understanding complex medical cases. Thus, advanced Generative AI (GenAI) techniques are increasingly being used to extract meaningful information from unstructured data, providing valuable information in predictive analytics and decision support systems.

Figure 1 depicts various types of EHR data, from medical tests, encompassing imaging scans, laboratory results, or genomic evaluations, to treatment details that can also include data from vital signs monitoring of the patient undergoing treatment.

As EHR data is continuously augmenting, AI is increasingly sought after in the healthcare sector as nearly 86% of organizations across multiple domains, including medicine and life science, are leveraging AI [3]. A recent study indicates that the global market for AI tools in healthcare was valued at 16.3 billion in 2022 and it is expected to reach nearly $174 billion by 2029. AI has the potential to reduce drug development costs by up to 70%, and hospital admission costs by up to 50%. There are currently 22% of hospitals and health systems only in the US that are in the early stages of implementing and integrating AI initiatives and tools, a number that will continue to increase in the coming years.

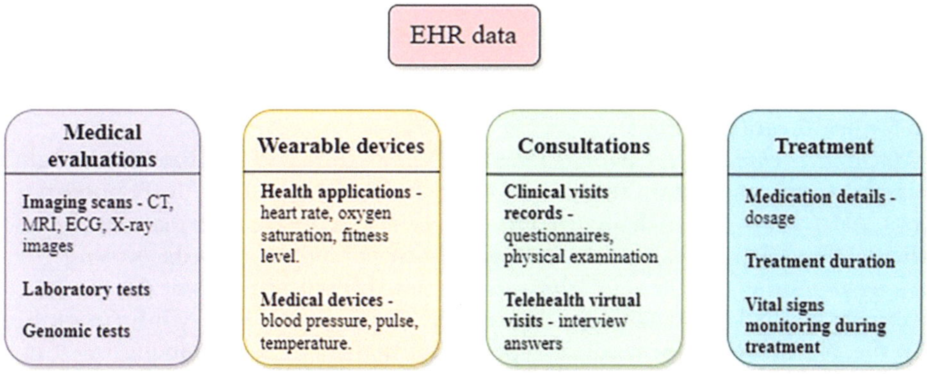

Figure 1.
Different types of EHR data.

Besides being able to enhance administrative workflows, AI in the healthcare sector can also be applied in key areas including health monitoring, medical imaging analysis and processing, clinical trials, disease detection and prevention, or diagnostics.

However, one of the most intriguing applications of AI in healthcare is represented by its role in EHR systems. In this regard, there are numerous benefits and innovative functionalities that AI can bring to EHR, such as:

2.1 Enhancing health data management and analysis

The management of healthcare data has always been a complicated and demanding task as there is an impressive amount of data and the traditional techniques frequently fall short of handling the "4 Vs" of big data: *velocity, volume, variety,* and *veracity*. Ensuring accurate recording, storage, and access of patient records, medical histories, treatment plans, or billing information imposes a significant strain on healthcare providers [4].

Several challenges that health data management faces include *the vast amount of data,* which is continuously growing at an exponential rate, being represented by structured (numeric values, coded information) and unstructured data (clinical notes, medical images, and signals). *Data accuracy* is critical in healthcare management as manual data entry and processing can be tough and can lead to errors, having a significant impact on patient care. Consequently, these inaccuracies may influence diagnostic errors, improper treatments, and billing inaccuracies, ultimately comprising the general quality of care. *Data integration* is another challenge that traditional methods struggle to address. As the healthcare systems and databases are often different and depend upon the healthcare providers, the processes of merging data into an exclusive standpoint are complex. This segmentation not only impedes streamlined data access and analysis but also results in delays in decision-making, potentially affecting patient care.

Furthermore, the lack of a comprehensive data overview can result in missed opportunities for identifying patterns and trends crucial for improving healthcare outcomes and operational efficiency. Advanced integration solutions are needed to address these issues and ensure seamless data flow across different platforms.

AI can have an enormous impact on significantly enhancing data accuracy and analysis in the healthcare sector. Conventional methods for data management frequently suffer from human errors, resulting in inaccurate patient records, diagnostic mistakes, and unsuitable treatments. Additionally, the traditional methods also fail to efficiently analyze the vast amounts of data generated in healthcare settings. In this regard, AI algorithms are adept at detecting and rectifying errors, assuring accurate data for healthcare providers.

Machine learning (ML) algorithms have the capability to analyze huge volumes of data, identifying certain trends and irregularities that may indicate potential errors. As an example, such algorithms can alert if there are unexplained abnormalities in a patient's health condition within their medical records. *Natural language processing (NLP)* algorithms, being considered a branch of AI, can be effectively applied in managing unstructured data, extracting key information from the doctors' notes or medical reports, standardizing it, and including it in the patient's structured files. These procedures may significantly reduce the likelihood of errors from manually entering the data, ensuring that all the essential information is accurately and systematically stored. NLP algorithms improve data processing by extracting context, identifying essential medical concepts, and accurately classifying data. This streamlines data entry enhances documentation accuracy and increases productivity for healthcare professionals. Furthermore, NLP enables advanced data analytics, helping healthcare providers retrieve specific insights from large datasets. This can lead to more informed decision-making processes and better quality of patient care services.

Moreover, having the capability to be continuously trained on multiple examples and scenarios and to adapt to real-life conditions, AI plays an essential part in enhancing EHR data management. As the volume of processed data is increasing, AI-based systems enhance their ability to identify potential errors, leading to more accurate data and reliable systems for improved patient outcomes and more streamlined healthcare operations.

2.2 Predictive analytics and personalization of patient care

Personalized medicine with AI: AI-driven EHR software is revolutionizing healthcare through the implementation of personalized medicine technologies that incorporate patient-specific data, such as medical history, lifestyle-related characteristics, genetic information, environmental factors, or treatment responses [5]. AI models can detect certain patterns by processing all these data, enabling accurate diagnoses and tailored treatments. This capability has led to the discovery of new biomarkers, genetic variations, and therapeutic targets [6], enabling more targeted and accurate treatments. An early predictive analysis allows healthcare providers to integrate preventative measures to reduce the impact of diseases and to enhance overall population health.

AI for remotely monitoring the healthcare of the patients to predict any abnormalities in their health status: When integrated with wearable devices and systems, the AI-powered EHR applications enable remote patient care through continuously gathering and analyzing real-time health data, such as vital signs, activity statuses, and medication adherence. This aspect empowers healthcare professionals to remotely detect any sign of impairment and to timely intervene in order to prevent medical complications or hospital readmissions. AI-powered remote healthcare monitoring significantly improves patient engagement, fostering the self-management of the patient's own health. One relevant use case to take into consideration is represented by Apple's wearable device, the Apple Watch, which has integrated AI algorithms designed to monitor heart health and predict potential abnormalities [7]. One of its key features is the detection of irregular heart rhythms, such as atrial fibrillation (AFib), which is a common and potentially serious heart condition. The watch utilizes a combination of optical heart sensors and advanced AI algorithms to continuously monitor heart rate data in real time. When the AI detects an irregular rhythm that might indicate AFib, the wearer receives an alert, prompting them to seek medical advice or take preventive measures. The data collected by the wearable device is also accessible to healthcare providers, enabling remote monitoring of the patient's heart health. Importantly, the Apple Watch's AFib detection algorithm has been certified by the U.S. Food and Drug Administration (FDA) as a Class II medical device [8]. This certification underscores the reliability and accuracy of the AI-driven monitoring system, ensuring that it meets rigorous safety and efficacy standards. As a result, healthcare providers and patients can have greater confidence in the early detection and management of heart conditions, potentially reducing the risk of severe outcomes.

Integrating AI into medical imaging and EHR systems: The volume of radiological imaging exams is rapidly increasing, leading to fatigue and burnout of radiologists who have to interpret an image every 3–4 seconds over an 8-hour workday. The rise of AI, especially deep learning models that can analyze vast amounts of imaging data with high accuracy and speed, has the potential to revolutionize medical diagnostics. These models, particularly convolutional neural networks (CNNs), excel at identifying patterns and anomalies in medical images that might be missed by the human eye. However, the effectiveness of CNNs or other AI-based models can be limited without the integration of clinical context, such as patient history, prior diagnoses or treatments, and laboratory results. Through the integration of these AI models with EHR systems, healthcare providers can benefit from a comprehensive view of a patient's medical history, enhancing the diagnostic process. These AI-powered imaging tools can continuously learn and improve from new data, leading to progressively more accurate and reliable results over time. This continuous progress helps in the early detection of diseases, accurate monitoring of disease progression, and the establishment of personalized treatment plans. AI-powered image recognition significantly boosts diagnostic accuracy, leading to improved patient outcomes, and elevating the overall quality of healthcare services. By swiftly and accurately identifying medical conditions, these advanced systems can support more precise diagnoses, enable timely interventions, and enhance the efficiency of healthcare delivery. Thus, this technology not only aids healthcare providers in making more informed decisions, but also provides personalized and effective care. Zebra Medical Vision has developed an AI-powered radiology platform that analyzes medical imaging, such as CT scans, X-rays, and MRIs, to detect a wide range of conditions, including fractures, lung nodules, and cardiovascular diseases. The AI algorithms are trained on millions of imaging data points, enabling them to provide highly accurate and rapid assessments. Once the AI analyzes the images, the results are automatically integrated into the patient's EHR system. This integration allows radiologists and other healthcare providers to access detailed AI-generated reports alongside the patient's medical history, lab results, and other relevant data within the EHR. The seamless integration facilitates quicker diagnosis and more informed decision-making, leading to more personalized and timely treatment plans [9].

Predictive analytics in terms of population health management: Predictive analytics integrate technologies that enhance individual and general population health and create more cost-efficient healthcare services. The approach leverages advanced data analysis methods, utilizing AI techniques to decode even the most intricate datasets. AI can pinpoint at-risk patients and facilitate proactive interventions to prevent or manage chronic diseases. Through the analysis of patient information, certain patterns and trends can be revealed in order to create personalized treatment plans. AI models can analyze population health data to detect disease trends, incidence, and patterns within communities, facilitating healthcare organizations to deploy targeted interventions and public health initiatives, thereby boosting health outcomes in terms of the general population.

2.3 The potential of using GenAI in revolutionizing EHR

It is widely recognized that GenAI will bring transformative changes to society with substantial potential to impact many industries. With the excitement surrounding GenAI, with ChatGPT now boasting over 100 million active users, there has been a growing interest in its potential application in the healthcare sector [10]. While

EHRs encompass a diverse range of structured information generated during regular clinical practice, such as vital signs, laboratory values, or clinical interventions, as well as unstructured data like medical narratives, discharge notes, or radiology documents, their full potential may remain unexplored due to significant challenges in data processing. Current strategies to overcome these challenges often depend on manual abstraction processes that may be time-consuming, complex, and susceptible to errors.

The domain of Large Language Models (LLMs) has emerged as groundbreaking technology, having the capability to revolutionize multiple aspects of healthcare management. In comparison to other NLP techniques, their potential to comprehend and generate human-like language has shown great promise in different medical tasks, such as summarization of clinical notes, medical decision support systems, or education and research [11]. When trained on large sets of EHRs (notes, laboratory values, treatment plans), LLMs offer the possibility for improved predictive performances in identifying potential health risks, forecasting disease progression, and developing personalized treatment strategies. This enhanced predictive capability can lead to earlier detection of conditions, more accurate prognoses, and more effective and tailored medical interventions, ultimately improving patient outcomes and healthcare efficiency.

The most common LLMs are developed by Google (Bidirectional Encoder Representations from Transformers—BERT [12]), OpenAI (Generative Pre-trained Transformer—GPT family [13], and Meta (Large Language Model Meta AI—Llama [14]).

LLMs can be applied in the management of EHR through the following:

- *Named entity recognition (NER)*: involves extracting structured data from unstructured information, identifying and classifying specific entities found in clinical texts or records. NER is mainly focused on two directions in EHR management: *extracting clinical terms* to understand medical history, diagnoses, treatments, and outcomes, thereby maintaining organized follow-up documentation; *extracting social determinants* (education, socio-demographic status, employment, or the level of access to healthcare services), which can drastically impact health outcomes of an individual. By integrating both medical and social backgrounds, healthcare providers can obtain a comprehensive understanding of a patient's health. BERT and GPT are currently used to perform NER tasks on EHR datasets while BioClinical-BERT and GPT4 were used for the extraction of social determinants from various EHR datasets, achieving good performances [15–18].

- *Information extraction*: extends beyond NER to include relation and event extraction, crucial for understanding sequences of medical events and building knowledge graphs. Various LLMs have been developed in order to facilitate better clinical decision-making. Several examples to mention are CancerBERT [19] that extracts phenotypic information from EHR data and NeighBERT that extracts relational data, improving the resolution of ambiguities in clinical texts [20].

- *Text summarization*: aims to efficiently collect the information and provide recommendations depending on the specific task. Text summarization can be applied to medical records or to the data collected from a single visit, offering information for both general and specialized healthcare needs. LLMs, with their advanced language understanding capacity, are highly efficient for text summarization, enabling the condensation of diverse textual structures into concise formats while preserving essential concepts and meanings. Summarizing patient medical information from medical histories, clinical notes, or laboratory results is a key application for LLMs as their implementation can turn the collected information into patient-friendly text, thus enhancing patient understanding and interaction with their healthcare providers [21, 22]. For example, Hartman et al. utilized fine-tuning for BERT and Bidirectional and Auto-Regressive Transformer (BART) models to produce discharge reports for neurology patients [21]. Moreover, LLMs have also been applied to summarize clinical progress notes, which may include a variety of laboratory results and documentation [23–25]. Therefore, in terms of clinical text summarization, it is proven that LLMs can surpass medical experts. Integrating them into clinical practice could reduce documentation burdens, enabling healthcare providers to focus more on patient care.

- *Text similarity*: entails identifying and measuring resemblance between multiple parts of clinical text data. LLMs frequently rely on semantic methods, which determine the similarity between texts, using semantic embeddings. The studies in this direction mainly focus on text matching, aiming at improving documentation of clinical notes or code standardization [26, 27].

- *Text classification*: involves classifying unstructured or semi-structured text data into particular categories that can significantly vary depending on the use case, encompassing medical characteristics (prescription orders, medication administration, drug usage, or adverse reactions), disorder types, or patient statuses. Several BERT-based models have been tested for the classification of medical aspects—mBERT, XLM-RoBERTa [28], of adverse drug reactions [29], or for the classification of disease coding from the International Classification of Diseases (ICD)—BioBERT, ClinicalXLNet, or Bio-ClinicalBERT [30, 31].

- *Conversational agents*: are transforming healthcare by providing interactive support to both patients and medical professionals. These LLMs-based systems are capable of handling a variety of tasks, from answering patient inquiries [32, 33] to assisting healthcare providers in clinical decision-making tasks [34]. In clinical settings, conversational agents that are based on high-performance LLMs (such as GPT-4o, or Llama-3) can assist doctors by providing summarizations of patient records, highlighting clinical information and even suggesting potential diagnoses to take into consideration [35].

Therefore, the potential of GenAI, particularly LLMs, in revolutionizing EHRs is profound and transformative. These advanced AI models offer significant enhancements in various EHR-related tasks, including NER, information extraction, text summarization, and classification, as well as conversational agents. By effectively processing and synthesizing structured, semi-structured, or unstructured data, LLMs improve diagnostic accuracy, streamline clinical workflows, and personalize patient care.

One example of using GenAI for revolutionizing how EHRs are managed is represented by Nuance Communications, a company developing AI-powered healthcare solutions, that has introduced the Dragon Ambient eXperience (DAX), a GenAI-powered tool designed to revolutionize the way Electronic Health Records (EHRs) are managed [36]. DAX uses advanced GenAI to automatically transcribe and summarize patient-physician interactions in real time, transforming these conversations into comprehensive clinical documentation. During a medical consultation, DAX listens to the conversation and generates detailed notes that accurately capture the nuances of the patient's symptoms, medical history, and treatment plan. These notes are then seamlessly integrated into the patient's EHR, significantly reducing the administrative burden on physicians. This allows healthcare providers to focus more on patient care rather than spending excessive time on documentation. Moreover, DAX's GenAI capabilities enable it to create personalized health summaries and actionable insights based on the conversation, which can help tailor treatment plans to the specific needs of the patient. The system also ensures that the documentation adheres to regulatory and billing requirements, enhancing the overall efficiency and accuracy of EHR management. In addition to this, Nuance Communications has also partnered with Microsoft to integrate GPT-4 to augment their platform, Dragon Ambient Intelligence. The aim is to generate notes from transcription minutes to further alleviate the administrative burdens on doctors [37].

LLMs enhance the ability to process and integrate diverse data types, making it possible to extract meaningful insights from vast amounts of medical information. This capability supports healthcare providers in understanding medical histories, diagnosing conditions, formulating treatment plans, and predicting patient outcomes with greater precision. The implementation of these technologies can significantly reduce the administrative burden on healthcare providers, improve patient engagement, and lead to more personalized and effective healthcare solutions.

As GenAI continues to advance, its integration into EHR systems promises to transform the landscape of healthcare. However, several challenges must be addressed, including ensuring data privacy and security, managing the heterogeneity of EHR systems, and improving the interpretability of AI models. Additionally, the high computational resources required for training and deploying LLMs can also be a barrier. Despite these challenges, the ongoing development and deployment of these technologies will play a crucial role in making healthcare more efficient, accurate, and patient-centered, ultimately improving health outcomes and the overall quality of care.

3. Critical issues surrounding the cybersecurity of EHRs

The integration of digital technologies in the healthcare sector has significantly enhanced the accessibility of medical services, with EHRs playing a pivotal role in this transformation. While EHRs are essential for modernizing healthcare delivery, they also create opportunities for cyber threats that can compromise patient data security and privacy. Check Point Software Technologies [38] highlights a significant trend, indicating that the healthcare domain faced an average of 1684 cyberattacks per week in the first quarter of 2023, with a 22% increase compared to the same period of the previous year. Consequently, the healthcare sector ranked as the third most targeted industry in 2023, surpassing finance, insurance, and communication.

Ransomware attacks were particularly prevalent, exploiting vulnerabilities in outdated systems and weak cybersecurity measures. These attacks [39] led to significant disruptions in patient care and financial losses, underscoring the urgent need for advanced cybersecurity solutions, such as AI-driven threat detection and robust encryption, to protect sensitive healthcare data and maintain system integrity.

The healthcare sector's cyber threats have not only increased in frequency but also in sophistication, with attackers increasingly targeting EHR systems with advanced techniques such as AI-powered malware [40]. The urgency for robust cybersecurity measures has never been greater, as these systems are now prime targets for global cybercrime.

This landscape not only jeopardizes patient data integrity and privacy, but also disrupts healthcare operations. It is essential for healthcare institutions to understand the types of cyber threats that may affect their systems and to devise and implement robust protective measures. Several key threats that can pose serious risks to the security and functionality of EHR systems are represented by:

- Phishing attacks: remain one of the most significant and deceptive cyber threats, where attackers utilize sophisticated methods to deceive healthcare providers into exposing sensitive information or unintentionally installing malicious software. Such attacks often pose as legitimate messages, such as emails from apparently trusted sources, making them particularly insidious. Healthcare providers might receive emails, enticing them to click on malicious websites or enter their login credentials, thereby assigning attackers unauthorized access to EHR systems. Phishing emails often exploit psychological manipulation tactics through creating a sense of urgency or fear, to prompt rapid actions from recipients. Successful phishing attempts can have serious repercussions, such as the release of ransomware that can seriously impair healthcare operations and grant unauthorized access to patient details. Recent reports emphasize an increase in phishing attacks directed toward the healthcare industry, underlining the necessity of effective digital security and continuous staff training to identify and lessen such risks [41].

- A recent report from the European Union Agency for Cybersecurity (ENISA) found that phishing attacks are one of the most common methods used by cybercriminals to break into healthcare systems. The report emphasizes the importance of implementing advanced email filtering systems, multi-factor authentication, and comprehensive cybersecurity education campaigns for healthcare personnel [42]. Furthermore, the American Hospital Association has reported a notable increase in phishing attacks, advocating for strengthened cybersecurity policies and cooperation with government authorities in order to enhance resilience against these ubiquitous threats [43].

- Malware and other cyber threats: infiltrate into healthcare networks through malicious email attachments, or susceptible software programs. Once entered, malware can propagate quickly, jeopardizing the confidentiality of patient data and compromising EHR systems. There are several malware infections that can impact EHR systems:

    - Ransomware: is a type of virus that encrypts information, particularly EHR data, until a ransom is paid. The consequences of these cyberattacks can be severe, leading to compromising patient care, significant financial losses, also harming the institution's reputation. In 2024, several major cyberattacks targeted healthcare systems, causing significant disruptions. One notable incident involved a ransomware attack on several Romanian hospitals which led to the temporary shutdown of critical systems, forcing the hospital to divert patients and cancel surgeries [44]. Another attack targeted a healthcare provider's EHR system, compromising sensitive patient data and leading to a costly recovery process. These incidents underscore the increasing frequency and severity of cyber threats in the healthcare sector [45]. For instance, the WannaCry ransomware attack in 2017 caused widespread disruption in the UK's National Health Service (NHS) [46], being estimated that £20 m was lost during the incident, and an additional £72 m as a result of IT support to restore systems and data.

- Trojan Horses: pose as trustworthy software in order to manipulate users into performing destructive things. These threats can leverage vulnerabilities to obtain private patient information or grant unauthorized access to healthcare networks. Several examples have been reported over time, causing widespread disruption across various sectors, including healthcare, such as Remcos-RAT or Agent Tesla [47].

- Botnets: can perform coordinated attacks on EHR systems, being represented by networks of compromised hardware. Among botnets, there are distributed denial-of-service (DDoS) attacks that can overburden healthcare systems and impair service accessibility. For instance, Passion Botnet has been applied in multiple DDoS attacks that targeted healthcare institutions across several countries (the US, Portugal, Spain, Germany, Poland, Finland, Norway, the Netherlands, and the UK). These attacks have created substantial disruptions by overwhelming healthcare servers with traffic, providing DDoS services on Telegram or other platforms. Thus, other cybercriminals were able to perform tailored attacks according to their requirements [48].

In addition to malware, other cyber threats that affect EHR systems may also include threats that arise from employees or contractors who intentionally or unintentionally compromise EHR data (*insider threats*) and that might be very difficult to identify and prevent; *supply chain attacks* that can be implemented by third-party suppliers or service providers through introducing security weaknesses into the systems.

Advancements in GenAI have enabled cybercriminals to execute highly sophisticated social engineering attacks by mimicking trusted individuals. With just a brief audio clip of an executive's voice, hackers can create convincing voice messages to manipulate employees into disclosing protected health information (PHI), wiring funds, or providing access credentials. Additionally, GenAI can craft more authentic-looking emails by replicating the writing style of trusted contacts. Beyond social engineering, even those with minimal coding skills can leverage GenAI to develop malware targeting EHR systems and other critical infrastructure [49].

The cybersecurity of EHR systems is a multifaceted and critical issue, encompassing comprehensive strategies to address a wide array of threats. Significant risks to patient data confidentiality, integrity, and other emerging threats are posed by phishing attacks, malware infections, botnets, or other developing threats. Additional threats such as medical device hacking, unsecured IoT devices, advanced ransomware techniques, insider threats, and supply chain vulnerabilities further complicate the security landscape.

Over the past 5 years, the nature of cyber threats targeting EHR systems has evolved from relatively simple phishing attacks to more complex, AI-driven ransomware and malware campaigns. This evolution reflects the increasing sophistication of cybercriminals and highlights the need for healthcare providers to continuously adapt their cybersecurity strategies.

In order to protect against these threats, healthcare institutions must implement robust security measures, such as advanced threat detection, periodic software updates, strict access limits, or regular employee training. Effective risk management procedures, in particular to what concerns third-party suppliers, are also necessary to protect against this type of cyberattack.

For digital healthcare systems to remain resilient and successful, it is imperative that these cybersecurity concerns are resolved. Healthcare organizations may effectively navigate the complex threat landscape and safeguard their digital assets by investing in advanced security technology, doing continuous research, and cultivating a cybersecurity awareness culture.

4. Tackling challenges in the integration of AI and cybersecurity in EHRs

Integrating AI with EHR has proven to offer significant advancements in healthcare, improving efficiency, accuracy, and patient outcomes. However, there are also complex cybersecurity challenges that must be addressed in order to protect EHR-sensitive data and guarantee the integrity of healthcare systems.

4.1 AI-related security concerns

The healthcare industry is also facing challenges related to the rise of sophisticated AI-powered cyberattacks. The integration of AI into EHR systems introduces security risks, such as adversarial attacks in which cybercriminals may manipulate input data to mislead AI models, leading to incorrect medical diagnoses or treatment recommendations [50]. For instance, by subtly altering input data, attackers can deceive AI models into making incorrect clinical decisions, such as misdiagnosing a condition or recommending inappropriate treatments [51]. Malicious AI attacks encompass several sophisticated techniques, such as data poisoning attacks (involves injecting malicious data into the training dataset, corrupting the AI model's learning process and potentially leading to unreliable outcomes), input manipulation attacks (manipulate the input data to deceive AI models, resulting in wrong predictions or classifications), membership inference attacks (determine whether a specific information was included in the training dataset, potentially exposing sensitive data, especially when a patient is a member of a sensitive category; e.g. cancer patient, association related to a particular sexual orientation), and model inversion attacks (infer sensitive information used during training by analyzing the model's outputs, exposing private information).

Clinical LLMs incorporated into EHR are considered an important tool in the advancements that have been made in healthcare analytics and clinical decision support systems (CDSS). Nevertheless, they are also susceptible to serious security risks, especially backdoor attacks that involve inserting malicious triggers, with modified labels to a particular target class, in a subset of training data [52]. Consequently, the backdoored model operates normally with clean inputs, but when it is triggered, it generates inaccurate outputs. This kind of attack undermines the model's integrity, which could result in inaccurate forecasts and potentially detrimental clinical judgments. The adoption of pre-tuned models from third-party services or public repositories for training data increases the severity of this vulnerability by providing attackers with various opportunities to introduce backdoored or hacked instances into the model.

Moreover, data poisoning and adversarial attacks are also particularly concerning, regarding Clinical LLMs, in numerous medical and critical-related applications, such as mortality prediction, where deliberate misclassifications using altered models may delay crucial medical procedures. These concerns seriously endanger patient safety in addition to undermining the validity of medical ML applications. For AI to be used in healthcare in a way that is both safe and efficient, there is an urgent need for focused research to detect and reduce these security threats.

4.2 Privacy and ethical concerns

Significant amounts of sensitive patient data are required to be used in order to integrate AI into EHR systems, which raises serious ethical and privacy issues. Ensuring compliance with regulations such as the Health Insurance Portability and Accountability Act (HIPAA) in the US and the General Data Protection Regulation (GDPR) in the EU is critical as they mandate rigorous data protection measures and patient consent protocols, making it more difficult to proceed with this kind of integration. HIPAA requires controlled access to PHI, detection of unauthorized changes, and PHI must also be encrypted both in transit (data traveling from a source to a destination; share PHI between two healthcare providers through cloud storage, emailing, or using other technologies) and at rest (storing electronic PHI—ePHI—on digital devices, such as servers, mobile devices, or hard drives) [53]. In a similar manner, GDPR enforces strict data privacy standards and it demands breach notifications within 72 hours, despite the severity of the incident [54]. Furthermore, AI systems must prevent reinforcing biases from the training datasets, as this could result in differences in patient treatment. A notable example is the underdiagnosis of diseases in minority populations due to a lack of diverse data in the training sets [55]. This highlights the importance of using comprehensive, representative data in AI training to ensure equitable treatment across all patient demographics. Implementing *fairness* and *bias reduction* techniques, along with thorough supervision, is necessary in order to address these ethical concerns. For example, using de-identified data and acquiring explicit patient consent are critical procedures to minimize confidentiality threats. By generating synthetic data, methods such as generative models can also lessen the need for actual patient data, improving privacy while preserving data utility.

The deployment of AI models in healthcare also arouses challenges about the utilization of patient data by other private entities. This could potentially raise privacy risks, even in cases in which data anonymization techniques are applied. The management of these risks depends on carefully constructed contracts that clearly define the parties' rights and responsibilities as well as who is responsible for any potential unfavorable results.

In addition to these regulations, the recently adopted EU AI Act [56] establishes a global standard for AI directives and emphasizes certain requirements on entities that are implementing, developing, or deploying AI systems within the EU. The Act creates rigorous compliance rules and divides AI systems into many risk categories, particularly for high-risk systems such as the ones used in healthcare that must adhere to strict data governance practices, continuous risk management, and transparency [57]. As an example, in the case of an AI system that directly communicates with individuals, the users must be made aware that they are interacting with an AI, thus guaranteeing transparency and responsibility. Additionally, post-market monitoring and incident notification are crucial when it comes to high-risk AI systems as malfunctions that potentially endanger people or interfere with critical infrastructures are required to be reported. By improving the security and dependability of AI systems in the healthcare industry, these measures aim to secure patient data and guarantee adherence to the new AI Act and GDPR laws.

4.3 Ensuring data integrity

Data integrity is a key aspect of ensuring the reliability and effectiveness of AI models when being integrated into EHR systems. The consistency and quality of the data that AI algorithms are trained on have a significant impact on their accuracy. Errors can be introduced by variations in data entry procedures throughout healthcare professionals, jeopardizing AI performance and resulting in inaccurate clinical judgments. Thus, upholding strict guidelines for data integrity is essential for the secure and efficient application of AI in healthcare environments. For example, if an AI system processes incomplete or corrupted data, it may generate flawed treatment recommendations, potentially compromising patient outcomes. This underscores the need for rigorous data validation and quality control mechanisms [58].

Several challenges that need to be addressed are the *variability* in data entry procedures throughout different healthcare providers that can lead to inconsistencies in data quality, considerably impacting AI models' performance. Incomplete or incorrectly entered data can lead to inaccurate AI predictions, which may result in misdiagnoses or improper treatment plans; *secure transmission and storage of data* as the transfer of sensitive patient information across various platforms and devices is essential within EHR systems, enhancing the risk for data breaches; *complexity* and *resource-intensive procedures* for robust encryption protocols; *cyberattacks* (ransomware and malware) can corrupt or encrypt patient-sensitive information, rendering it inaccessible; management of data versioning and updates throughout different systems and versions is critical as the EHR systems continuously evolve and it is a complex process to maintain accurate and up-to-date patient records.

4.4 Interoperability and integration challenges

Interoperability and integration issues are major when integrating AI with current EHR systems because a large number of healthcare providers still rely on outdated infrastructures and programs that probably may not support the sophisticated requirements of AI technologies. Considering the considerable financial investments and logistical complexities together with the cybersecurity measures that have to be applied, there is a significant barrier to seamless AI integration into EHR systems.

Different EHR systems from various healthcare providers use diverse data structures and formats, posing the lack of standardization as a challenge in EHR data formats. Such discrepancies may lead to data processing errors and shortcomings, eventually affecting the performance and reliability of the applied AI models. Moreover, ensuring secure and efficient data exchange between EHR systems and AI applications is essential for maintaining data integrity and confidentiality of patient records. This includes implementing secure Application Programming Interfaces (APIs) and data transmission procedures that prevent unauthorized access and data breaches. As compliance with regulatory requirements needs to be assured, another level of complexity is added to the integration process.

5. Optimizing healthcare systems through AI and cybersecurity

Imagine a healthcare system where administrative burdens vanish, patient care becomes personalized and accurate, and data breaches are relics of the past. This vision is starting to become a reality as more and more healthcare providers have started to integrate AI models and systems into their operations, taking also into consideration the important aspect of cybersecurity. However, achieving such a transformation is not without challenges. From ensuring data privacy and integrity to overcoming interoperability issues and tackling ethical questions, the path to an optimized healthcare system is complex. AI together with cybersecurity measures has the potential to revolutionize healthcare by addressing these obstacles, ensuring a secure and effective future for EHR systems.

AI technology can significantly improve the *operational efficiency of healthcare systems* by automating routine tasks. AI-driven solutions can automate the documentation of health data, streamline patient scheduling, and manage billing procedures. For instance, Epic is the most widely used EHR system worldwide and it is ranked in the top 5 EHR providers. It is currently using GenAI, such as GPT-4, in its EHR systems to enable functionalities like drafting personalized responses to patient messages, collecting data from prior nursing shifts, generating hand-off reports, and using ambient listening during medical visits of patients, proving that there are a lot of potential benefits of integrating AI when it comes to interoperability and standardization [59].

AI tools can also predict patient admission rates, optimize staff allocation, and efficiently manage inventories. Predictive analytics can be applied to foresee patient surges, enabling healthcare providers to adequately prepare and assure that resources are allocated where they are needed most. This level of operational planning can significantly diminish wait times, improve the patient experience, and enhance the overall quality of care.

The AI's ability to quickly and precisely analyze and process large amounts of data may lead to *improved patient outcomes*. AI-enabled CDSS can assist healthcare professionals in diagnosing conditions, suggesting therapies, and spotting potentially harmful drug combinations. AI can identify patterns and correlations that might be overlooked by the human eye, resulting in early disease detection or more tailored treatment options. For example, AI tools can analyze imaging data to identify certain anomalies at an early stage, thus increasing the likelihood that a therapy would be effective [60]. AI is also capable of real-time patient data monitoring, which not only allows for prompt actions but also notifies healthcare personnel of significant changes in a patient's health status during the day [61].

Additionally, as healthcare systems become increasingly digitized, the importance of cybersecurity cannot be overstated. Protecting sensitive patient data from cyber threats is crucial to maintaining trust and ensuring the integrity of healthcare operations. The integration of advanced cybersecurity measures, such as *encryption*, *multi-factor authentication*, and *continuous monitoring*, is essential to safeguard EHR systems against unauthorized access and data breaches [62]. Encryption ensures that patient data is secure during transmission and storage, while multi-factor authentication adds an extra layer of security by requiring additional verification steps beyond just passwords. Continuous monitoring systems can detect and respond to cyber threats in real time, helping to mitigate potential damage before it escalates.

AI plays a pivotal role in strengthening cybersecurity by *identifying and responding to threats in real time*. ML algorithms can detect unusual patterns of behavior that may indicate cyberattacks, such as data exfiltration or unauthorized access attempts. For example, AI-driven security systems can analyze vast amounts of network traffic and user activity data to identify potential threats before they cause significant damage. This proactive approach allows healthcare organizations to defend against cyber threats more effectively and protect sensitive patient information. AI-driven enhancements in cybersecurity include automating incident response. AI virtual assistants can quickly isolate impacted systems and cut off attacker access, accelerating the response time and allowing security teams to focus on more complex threats. Implementing a cybersecurity mesh architecture, which allows for modular security functions to interoperate through supportive layers, enhances the flexibility and scalability of security measures.

One use case that is worth mentioning is Cleveland Clinic, which aims to integrate AI with cybersecurity strategies to enhance the protection of its EHR systems. They have developed an advanced digital strategy that includes AI-driven analytics and cloud technologies, improving caregiver efficiency while maintaining a strong cybersecurity posture. The integration allows the clinic to utilize years of EHR data more effectively while protecting it from cyber threats through AI-enhanced threat detection and risk management [63]. The implementation of AI and cybersecurity at Cleveland Clinic has led to measurable improvements in operational efficiency and patient outcomes. By utilizing AI-driven analytics and real-time threat detection, the clinic has significantly reduced the time required for data processing and enhanced its ability to preemptively address potential cyber threats, thereby minimizing disruptions to patient care.

Consolidating key security functions on single platforms enables healthcare organizations to reduce administrative burdens and increase the efficiency of their security operations. An AI-based approach also facilitates the integration of new security technologies as they emerge, ensuring that the system remains up-to-date with the latest advancements.

6. Future visions for secure EHR systems

As the digital landscape of healthcare continues to evolve, the future of EHR systems promises unprecedented advancements in both accessibility and security. The seamless integration of AI, blockchain, and quantum cryptography is poised to redefine how patient data is managed, ensuring it remains both highly accessible to authorized users and impervious to cyber threats. However, navigating this future demands a delicate balance between innovation and stringent security measures, ethical considerations, and regulatory compliance. As we look ahead, the challenge lies in leveraging these cutting-edge technologies to build EHR systems that are not only advanced and efficient but also secure and trustworthy, laying the groundwork for a new era in healthcare.

Advanced AI techniques: Future EHR systems will likely leverage AI for predictive analytics, personalized medicine, and improved clinical decision support. However, with these advancements comes the need for robust security measures to protect against sophisticated cyber threats. AI-driven security solutions that can detect and respond to anomalies in real time will be crucial in safeguarding EHR data. For instance, ML algorithms can continuously monitor network traffic and user behavior to identify potential security breaches before they occur, thereby providing a proactive defense mechanism against cyberattacks.

Blockchain technology: By providing a decentralized and immutable ledger for health records, blockchain can ensure that patient data is tamper-proof and verifiable, enhancing the security and integrity of EHR systems. Each transaction or modification in the blockchain is recorded and cannot be altered retroactively, thus maintaining the integrity of health records. Moreover, blockchain can facilitate secure data sharing among healthcare providers, patients, and other stakeholders while ensuring compliance with the regulations of HIPAA and GDPR.

Quantum computing and cryptography: Quantum computers could break traditional encryption methods, necessitating the development of quantum-resistant cryptographic techniques with the potential to both revolutionize and challenge current cybersecurity frameworks. Future EHR systems will need to incorporate these advanced cryptographic solutions to protect patient data against the computational power of quantum machines. Research into post-quantum cryptography is essential to prepare for this impending shift and to ensure that EHR systems remain secure in a quantum future.

Enhanced interoperability and data exchange: Standardizing data formats and protocols will facilitate the integration of disparate systems and enhance the flow of information across the healthcare ecosystem. Secure APIs and encrypted data transmission methods will be essential to protect data during exchange. Additionally, advancements in federated learning can enable collaborative AI model training across institutions without sharing raw data, thereby preserving patient privacy while leveraging collective insights.

Patient-centric models and data ownership: Empowering patients with control over their health data is a critical aspect of future EHR systems. Patient-centric models will allow individuals to own their health records, granting access to healthcare providers as needed. This approach not only enhances patient engagement and trust but also ensures that data is shared securely and transparently. Implementing secure

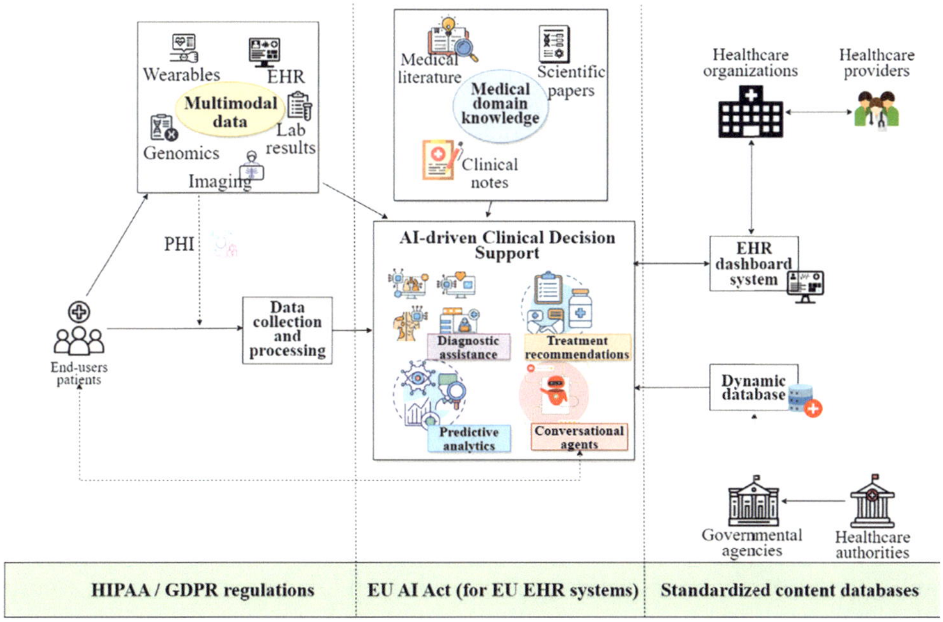

Figure 2.
Proposed conceptual framework for a future AI-driven CDSS.

patient portals and mobile applications with robust authentication mechanisms will be crucial for facilitating this model of data ownership.

In **Figure 2**, we propose a conceptual framework for a future vision of a secured AI-enabled EHR system.

The figure outlines a comprehensive framework for a future AI-driven CDSS in healthcare. It integrates multimodal data sources including wearables, EHR, laboratory results, genomics, and medical imaging together with medical domain knowledge from scientific papers, clinical notes, and literature. Data from end-users (patients) is collected and processed, adhering to HIPAA/GDPR regulations, to support diagnostic assistance, treatment recommendations, predictive analytics, and conversational agents within the AI-driven CDSS. Conversational agents will be based on multimodal LLMs in order to support the diagnosis for the healthcare provider as well as to have controlled communication with the patients. Outputs from the CDSS are utilized in EHR dashboard systems for healthcare providers and organizations, as well as dynamic databases interfacing with governmental agencies and healthcare authorities. The framework ensures compliance with the EU AI Act for EHR systems and utilizes standardized content databases for consistent data interoperability and exchange.

7. Conclusions

The integration of AI and advanced cybersecurity measures into EHR systems marks a significant step forward in optimizing healthcare operations. Throughout this chapter, we have explored how AI can improve the healthcare system, especially, the challenges that cybersecurity imposes in the healthcare sector, how AI and cybersecurity can provide both challenges and opportunities in this domain, as well as a conceptual framework for a future vision of an AI-based CDSS.

AI-driven tools are revolutionizing healthcare by automating routine tasks, aiding in clinical decision-making, and providing real-time threat detection. These advancements not only improve operational efficiency but also enhance patient outcomes by enabling personalized and timely medical interventions. For instance, predictive analytics and AI algorithms analyze patient data to forecast health trends, allowing for early interventions that can significantly improve patient care. Implementing robust cybersecurity frameworks, such as blockchain technology, ensures that patient data remains secure and intact. These technologies provide multiple layers of security, continuous verification, and immutable records, which are critical in preventing data breaches and maintaining the integrity of healthcare systems.

Addressing cybersecurity challenges through AI-driven enhancements, such as automated incident response and a cybersecurity mesh architecture, further strengthens the healthcare infrastructure. By consolidating key security functions and reducing administrative burdens, healthcare organizations can achieve a flexible and scalable security posture that adapts to emerging threats.

Moreover, compliance with data privacy regulations like HIPAA and GDPR is paramount in protecting patient data and maintaining trust. Regular audits, stringent security measures, and staff education ensure that healthcare practices align with regulatory standards, safeguarding patient privacy and fostering a culture of security within the organization.

In summary, the fusion of AI and cybersecurity within EHR systems offers a transformative potential for healthcare. By enhancing operational efficiency, improving patient outcomes, and ensuring robust data security, these technologies pave the way for a more resilient and effective healthcare system. Continuous innovation and strategic investment in AI and cybersecurity are essential for navigating the complexities of modern healthcare, ultimately ensuring a safer and more efficient future for patient care.

Acknowledgements

The authors acknowledge the use of ChatGPT for language polishing of the manuscript.

Author details

Elena-Anca Paraschiv*, Carmen Elena Cîrnu and Adrian Victor Vevera
National Institute for Research and Development in Informatics – ICI Bucharest, Bucharest, Romania

*Address all correspondence to: elena.paraschiv@ici.ro

References

[1] Li L, Zhou J, Gao Z, Hua W, Fan L, Yu H, et al. A scoping review of using large language models (LLMs) to investigate electronic health records (EHRs) [Internet]. arXiv. 2024. Available from: https://arxiv.org/abs/2405.03066 [Accessed: July 31, 2024]

[2] Cox D. Working with Semi-Structured & Unstructured Data in Healthcare [Internet]. Hakkoda; 2024. Available from: https://hakkoda.io/resources/unstructured-data-in-healthcare/ [Accessed: July 31, 2024]

[3] Siwicki B. 86% of Healthcare Companies Use some Form of AI [Internet]. Healthcare IT News; 2017. Available from: https://www.healthcareitnews.com/news/86-healthcare-companies-use-some-form-ai

[4] How AI is Enhancing Healthcare Data Management. Thoughtful [Internet]. Available from: https://www.thoughtful.ai/blog/how-ai-is-enhancing-healthcare-data-management

[5] Albanians L, How AI. Will Enhance Healthcare EMR/EHR Software In 2024? [Internet]. Medium; 2024. Available from: https://medium.com/@Larisa10/how-ai-will-enhance-healthcare-emr-ehr-software-in-2024-201e8465cbc9

[6] The Role of Artificial Intelligence in Personalized Medicine - Biotech Spain [Internet]. Available from: https://biotech-spain.com/en/articles/the-role-of-artificial-intelligence-in-personalized-medicine/

[7] Heart Health Notifications on your Apple Watch [Internet]. 2024. Available from: https://support.apple.com/en-us/120276 [Accessed: August 26, 2024]

[8] The FDA Approves Apple Watch's Heart Monitoring Tool for Use in Clinical Trials [Internet]. 2024. Available from: https://www.mobihealthnews.com/news/fda-approves-apple-watchs-heart-monitoring-tool-use-clinical-trials [Accessed: August 26, 2024]

[9] Zebra Healthcare Technology Solutions [Internet]. 2024. Available from: https://www.zebra.com/gb/en/solutions/industry/healthcare.html [Accessed: August 26, 2024]

[10] Raza MM, Venkatesh KP, Kvedar JC. Generative AI and large language models in health care: Pathways to implementation. npj Digital Medicine. 2024;**7**(1):62

[11] Thirunavukarasu AJ, Ting DSJ, Elangovan K, et al. Large language models in medicine. Nature Medicine. 2023;**29**:1930-1940. DOI: 10.1038/s41591-023-02448-8

[12] Devlin J, Chang MW, Lee K, Toutanova K. BERT: Pre-training of deep bidirectional transformers for language understanding [internet]. arXiv. 2018. Available from: https://arxiv.org/abs/1810.04805

[13] Ye J, Chen X, Xu N, Shao Z, Liu S, Cui Y, et al. A comprehensive capability analysis of GPT-3 and GPT-3.5 series models [Internet]. arXiv. 2023. Available from: https://arxiv.org/pdf/2303.10420 [Accessed: May 8, 2024]

[14] Touvron H, Lavril T, Izacard G, Martinet X, Lachaux MA, Lacroix T, et al. LLaMA: Open and efficient foundation language models. arXiv [Internet]. 2023. Available from: https://arxiv.org/abs/2302.13971

[15] Jitesh BN, Mittal A, He Z, Luo X. Zero-shot learning with minimum instruction to extract social determinants and family history from clinical notes using GPT model. 2023. arXiv [cs. CL]. Available from: http://arxiv.org/abs/2309.05475

[16] Guevara M, Chen S, Thomas S, Chaunzwa TL, Franco I, Kann BH, et al. Large language models to identify social determinants of health in electronic health records. npj Digital Medicine [Internet]. 2024;**7**(1):1-14. Available from: https://www.nature.com/articles/s41746-023-00970-0 [Accessed: February 1, 2024]

[17] Richie R, Ruiz VM, Han S, Shi L, Tsui FR. Extracting social determinants of health events with transformer-based multitask, multilabel named entity recognition. Journal of the American Medical Informatics Association. 2023;**30**(8):1379-1388. DOI: 10.1093/jamia/ocad046

[18] Wang X, Gupta D, Killian M, He Z. Benchmarking transformer-based models for identifying social determinants of health in clinical notes. IEEE International Conference on Healthcare Informatics. 2023;**2023**:570-574. DOI: 10.1109/ichi57859.2023.00102. Epub 2023 Dec 11

[19] Zhou S, Wang N, Wang L, Liu H, Zhang R. CancerBERT: A cancer domain-specific language model for extracting breast cancer phenotypes from electronic health records. Journal of the American Medical Informatics Association. 2022;**29**(7):1208-1216. DOI: 10.1093/jamia/ocac040

[20] Singh A, Krishnamoorthy S, Ortega JE. NeighBERT: Medical entity linking using relation-induced dense retrieval. Journal of Healthcare Informatics Research. 2024;**8**(2):353-369

[21] Hartman VC, Bapat SS, Weiner MG, Navi BB, Sholle ET, Campion TR Jr. A method to automate the discharge summary hospital course for neurology patients. Journal of the American Medical Informatics Association. 2023;**30**(12):1995-2003. DOI: 10.1093/jamia/ocad177

[22] Pal K, Bahrainian SA, Mercurio L, Eickhoff C. Neural summarization of electronic health records [Internet]. arXiv. 2023. Available from: https://arxiv.org/abs/2305.15222 [Accessed: July 31, 2024]

[23] Van Veen D, Van Uden C, Blankemeier L, Delbrouck JB, Aali A, Bluethgen C, et al. Adapted large language models can outperform medical experts in clinical text summarization. Nature Medicine. 2024;**30**(4):1134-1142. DOI: 10.1038/s41591-024-02855-5. Epub 2024 Feb 27

[24] Chuang YN, Tang R, Jiang X, Hu X. SPeC: A soft prompt-based calibration on performance variability of large language model in clinical notes summarization [Internet]. arXiv. 2023. Available from: https://arxiv.org/abs/2303.13035 [Accessed: July 31, 2024]

[25] Tung JYM, Gill SR, Sng GGR, Lim DYZ, Ke Y, Tan TF, et al. Comparison of the quality of discharge letters written by large language models and junior clinicians: Single-blinded study. Journal of Medical Internet Research. 2024;**26**:e57721. DOI: 10.2196/57721

[26] Xiong Y, Chen S, Chen Q, Yan J, Tang B. Using character-level and entity-level representations to enhance bidirectional encoder representation from transformers-based clinical semantic textual similarity model: ClinicalSTS modeling study. JMIR Medical Informatics. 2020;**8**(12):e23357. DOI: 10.2196/23357

[27] Wang L, Bi W, Zhao S, Ma Y, Lv L, Meng C, et al. Investigating the impact of prompt engineering on the performance of large language models for standardizing obstetric diagnosis text: Comparative study. JMIR Formative Research. 2024;**8**:e53216. DOI: 10.2196/53216

[28] Zhang X, Song X, Feng A, Gao Z. Multi-self-attention for aspect category detection and biomedical multilabel text classification with BERT. Mathematical Problems in Engineering. 2021;**2021**:1-6

[29] Chaichulee S, Promchai C, Kaewkomon T, Kongkamol C, Ingviya T, Sangsupawanich P. Multi-label classification of symptom terms from free-text bilingual adverse drug reaction reports using natural language processing. PLoS One. 2022;**17**(8):e0270595

[30] Chen PF, Chen KC, Liao WC, Lai F, He TL, Lin SC, et al. Automatic international classification of diseases coding system: Deep contextualized language model with rule-based approaches. JMIR Medical Informatics [Internet]. 2022;**10**(6):e37557. Available from: https://medinform.jmir.org/2022/6/e37557 [Accessed: May 27, 2023]

[31] Aden I, Christopher CC, Reyes-Aldasoro. International classification of diseases prediction from MIMIIC-III clinical text using pre-trained ClinicalBERT and NLP deep learning models achieving state of the art. Big Data and Cognitive Computing. 2024;**8**(5):47-47

[32] Garcia P, Ma SP, Shah S, Smith M, Jeong Y, Devon-Sand A, et al. Artificial intelligence–generated draft replies to patient inbox messages. JAMA Network Open. 2024;**7**(3):e243201-e243201

[33] Zhang Z, Yao Z, Zhou H, Ouyang F, Yu H. EHRTutor: Enhancing patient understanding of discharge instructions. 2023. arXiv [cs.CL]. Available from: http://arxiv.org/abs/2310.19212

[34] Li Y, Li Z, Zhang K, Dan R, Jiang S, Zhang Y. ChatDoctor: A medical chat model fine-tuned on a large language model meta-AI (LLaMA) using medical domain knowledge. Cureus. 2023;**15**(6):e40895. DOI: 10.7759/cureus.40895

[35] Hirosawa T, Harada Y, Yokose M, Sakamoto T, Kawamura R, Shimizu T. Diagnostic accuracy of differential-diagnosis lists generated by generative pretrained transformer 3 Chatbot for clinical vignettes with common chief complaints: A pilot study. International Journal of Environmental Research and Public Health. 2023;**20**(4):3378

[36] DAX Copilot [Internet]. 2024. Available from: https://www.nuance.com/healthcare/dragon-ai-clinical-solutions/dax-copilot.html [Accessed: August 26, 2024]

[37] Microsoft Adds GPT-4 Generative AI to Nuance's Dragon Medical Voice Assistant [Internet]. 2023. Available from: https://voicebot.ai/2023/03/20/microsoft-adds-gpt-4-generative-ai-to-nuances-dragon-medical-voice-assistant/ [Accessed: August 26, 2024]

[38] Check Point Research. Global Cyberattacks Continue to Rise with Africa and APAC Suffering Most [Internet]. Check Point Blog; 2023. Available from: https://blog.checkpoint.com/research/global-cyberattacks-continue-to-rise/

[39] Biggest Cyber Attacks, Data Breaches Ransomware Attacks: February 2024 [Internet]. 2024. Available from: https://www.cm-alliance.com/cybersecurity-blog/

biggest-cyber-attacks-data-breaches-ransomware-attacks-february-2024 [Accessed: August 26, 2024]

[40] 2024 Healthcare Cybersecurity Predictions [Internet]. 2024. Available from: https://www.linkedin.com/pulse/2024-healthcare-cybersecurity-predictions-cylera-mbmbc/ [Accessed: August 26, 2024]

[41] Greenberg I. A Rise in Healthcare Cyber Attacks Calls for Zero Trust [Internet]. World Economic Forum; 2023. Available from: https://www.weforum.org/agenda/2023/05/cyber-attacks-on-healthcare-rise-zero-trust/

[42] ENISA. ENISA Threat Landscape 2023 [Internet]. ENISA; 2023. Available from: https://www.enisa.europa.eu/publications/enisa-threat-landscape-2023

[43] dsamuels_drupal. Keeping Hospitals and Patients Safe against Cyberattacks. AHA News [Internet]. Available from: https://www.aha.org/news/perspective/2023-02-03-keeping-hospitals-and-patients-safe-against-cyberattacks

[44] Over a Hundred Romanian Hospitals Affected by Ransomware Attack [Internet]. 2024. Available from: https://www.digitalhealth.net/2024/02/over-a-hundred-romanian-hospitals-affected-by-ransomware-attack/ [Accessed: August 26, 2024]

[45] The Major Cyberattacks that have Affected Healthcare Systems in 2024 [Internet]. 2024. Available from: https://www.tebra.com/theintake/practice-operations/medical-news/the-major-cyberattacks-that-have-affected-healthcare-systems-in-2024 [Accessed: August 26, 2024]

[46] Stevens J. The WannaCry Attack and the NHS [Internet]. Emerge Digital; 2019. Available from: https://emerge.digital/resources/the-wannacry-attack-and-the-nhs [Accessed: July 31, 2024]

[47] Current Malware Threats Targeting the Healthcare and Public Health (HPH) Sector. AHA [Internet]. Available from: https://www.aha.org/other-cybersecurity-reports/2020-06-17-current-malware-threats-targeting-healthcare-and-public

[48] Davis J. Passion Botnet Cyberattacks Hit Healthcare, as Actors Offer Threat as DDoS-as-a-Service [Internet]. SC Media; 2023. Available from: https://www.scmagazine.com/analysis/passion-botnet-cyberattacks-hit-healthcare-as-actors-offer-threat-as-ddos-as-a-service [Accessed: July 31, 2024]

[49] Healthcare Security in 2024: The Cyberthreat Landscape [Internet]. 2024. Available from: https://www.bdo.com/insights/industries/healthcare/healthcare-security-in-2024-the-cyberthreat-landscape [Accessed: August 26, 2024]

[50] Hague M. Artificial Intelligence Poses Cybersecurity Risks in Healthcare [Internet]. Meditology Services; 2023. Available from: https://www.meditologyservices.com/artificial-intelligence-poses-cybersecurity-risks-in-healthcare/

[51] Sorin V, Soffer S, Glicksberg BS, Barash Y, Konen E, Klang E. Adversarial attacks in radiology – A systematic review. European Journal of Radiology. 2023;**167**:111085. DOI: 10.1016/j.ejrad.2023.111085

[52] Lyu W, Bi Z, Wang F, Chen C. BadCLM: Backdoor attack in clinical language models for electronic health records [Internet]. arXiv. 2024. Available from: https://arxiv.org/abs/2407.05213 [Accessed: July 31, 2024]

[53] Lau V. HIPAA Encryption: Requirements, Best Practices & Software [Internet]. Kiteworks. Your Private Content Network; 2023. Available from: https://www.kiteworks.com/hipaa-compliance/hipaa-encryption [Accessed: July 31, 2024]

[54] Guidelines 9/2022 on Personal Data Breach Notification Under GDPR. European Data Protection Board [Internet]. Available from: https://www.edpb.europa.eu/our-work-tools/our-documents/guidelines/guidelines-92022-personal-data-breach-notification-under_en

[55] Algorithms Are Making Decisions About Health Care, Which May Only Worsen Medical Racism [Internet]. 2022. Available from: https://www.aclu.org/news/privacy-technology/algorithms-in-health-care-may-worsen-medical-racism [Accessed: August 25, 2024]

[56] European Parliament. EU AI Act: First Regulation on Artificial Intelligence [Internet]. European Parliament; 2023. Available from: https://www.europarl.europa.eu/topics/en/article/20230601STO93804/eu-ai-act-first-regulation-on-artificial-intelligence

[57] The New EU AI Act – the 10 Key Things you Need to Know Now [Internet]. 2023. Available from: https://www.dentons.com/en/insights/articles/2023/december/14/the-new-eu-ai-act-the-10-key-things-you-need-to-know-now

[58] Data Quality and Data Privacy Challenges in AI Applications [Internet]. 2024. Available from: https://www.researchgate.net/publication/378904709_Data_quality_and_data_privacy_challenges_in_AI_applications [Accessed: August 26, 2024]

[59] Siwicki B. How Epic Is Using AI to Change the Way EHRs Work [Internet]. Healthcare IT News; 2023. Available from: https://www.healthcareitnews.com/news/how-epic-using-ai-change-way-ehrs-work

[60] Silva HECD, Santos GNM, Leite AF, et al. The use of artificial intelligence tools in cancer detection compared to the traditional diagnostic imaging methods: An overview of the systematic reviews. PLoS One. 2023;**18**(10):e0292063. DOI: 10.1371/journal.pone.0292063

[61] Shaik T, Tao X, Higgins N, Li L, Gururajan R, Zhou X, et al. Remote patient monitoring using artificial intelligence: Current state, applications, and challenges. WIREs Data Mining and Knowledge Discovery [Internet]. 2023;**13**(2):e1485. Available from: https://wires.onlinelibrary.wiley.com/doi/full/10.1002/widm.1485

[62] Cobrado UN, Sharief S, Regahal NG, Zepka E, Mamauag M, Velasco LC. Access control solutions in electronic health record systems: A systematic review. Informatics in Medicine Unlocked. 2024;**49**:101552

[63] The Transformative Power of AI in Healthcare [Internet]. 2024. Available from: https://digitalsalutem.com/the-transformative-power-of-ai-in-healthcare/ [Accessed: August 26, 2024]

Chapter 3

Perspective Chapter: Securing the Future of Digital Healthcare with EHR Systems for Modern Cyber Threats

Ritviksiddha Penchala and Venkata Krishna Parimala

Abstract

This chapter explores the growing role of Electronic Health Record (EHR) systems in healthcare and the cybersecurity challenges they face. It discusses how EHR systems enhance the accessibility, coordination, and analysis of patient data, improving clinical decision-making and public health initiatives. However, it also emphasizes the vulnerabilities these systems face, particularly in the face of cyberattacks such as ransomware, phishing, and data breaches. The chapter examines existing security measures such as encryption, access control, and multi-factor authentication, while also considering the potential of newer technologies such as artificial intelligence, machine learning, and blockchain to further safeguard these systems. It calls for continuous improvement and innovation to ensure EHR systems remain secure and reliable as healthcare increasingly relies on digital infrastructure.

Keywords: electronic health records, cybersecurity, privacy, health informatics, data breaches, ransomware, artificial intelligence, Blockchain

1. Introduction

The widespread adoption of Electronic Health Record (EHR) systems has revolutionized healthcare, enhancing the accessibility, analysis, and accuracy of patient data. Physicians, researchers, and data analysts can now derive insights that significantly impact both individual and public health outcomes. However, this unprecedented storage and usage of sensitive health information also introduces substantial cybersecurity risks. The security of EHR systems is paramount for patients, healthcare providers, and hospital networks, as cyber threats targeting healthcare data continue to escalate.

Existing literature highlights both the strengths and weaknesses of current EHR security measures. Standard approaches such as data encryption, access control mechanisms, and multi-factor authentication are generally effective but come with high computational and administrative costs [1]. Moreover, evolving cyber threats, including ransomware attacks, phishing schemes, and insider threats, challenges the resilience of these measures [2, 3].

Recent advancements propose novel solutions to strengthen EHR security. Blockchain technology has been explored as a means of enhancing data integrity and decentralization, reducing the risk of single points of failure [4]. However, its implementation presents computational challenges and scalability concerns. Artificial intelligence (AI) and machine learning techniques are also emerging as powerful tools for real-time threat detection and anomaly recognition in EHR networks [5]. These technologies enable proactive identification and mitigation of potential security breaches, reducing response times and enhancing overall system resilience.

Additionally, organizations such as the National Institute of Standards and Technology (NIST) provide guidelines for improving cybersecurity in healthcare IT, emphasizing risk assessment, privacy-preserving frameworks, and continuous monitoring [6]. Despite these advancements, the healthcare industry remains a primary target for cybercriminals due to the high value of medical records on the black market and the reliance on legacy systems that lack modern security protections [7].

Given the increasing reliance on EHR networks for medical information storage and sharing, it is crucial to explore cutting-edge security measures to address these evolving threats. This chapter examines current security mechanisms, emerging technologies, and their limitations, aiming to identify the most effective strategies for ensuring that EHR systems remain both secure and technologically advanced.

2. Current EHR systems

Electronic Health Record (EHR) systems are a cornerstone of modern healthcare, facilitating the digitization of patient information for improved storage, accessibility, and analysis. These systems enhance healthcare providers' ability to deliver accurate, efficient, and secure patient care while enabling seamless coordination across hospital networks. However, despite their advantages, current EHR systems face significant challenges, including cybersecurity threats, usability issues, interoperability limitations, and reliance on outdated legacy infrastructure. These shortcomings impact the security, efficiency, and overall effectiveness of EHR systems. Before examining these challenges and ongoing improvements, it is essential to first explore the current EHR landscape and the factors shaping its architecture.

2.1 Objectives

Current EHR systems are designed with several critical objectives at their core, each of which enhances the overall healthcare delivery process. These objectives include improving accessibility and availability of patient information, enhancing coordination among healthcare providers, optimizing clinical decision-making, increasing administrative efficiency, and facilitating research and public health initiatives through accessible, anonymized data.

One of the primary benefits of EHR systems is their ability to provide healthcare professionals with instantaneous access to comprehensive and detailed patient records. This accessibility is crucial in time-sensitive medical situations, such as emergencies, where real-time data retrieval can significantly impact patient outcomes. Studies indicate that EHRs have streamlined access to medication records, lab results, vaccination history, imaging studies, and prior diagnoses, thereby reducing redundancy and improving care quality [8, 9]. The digitalization of health records has led to a significant reduction in medical errors and improved patient safety.

EHR systems enable seamless coordination among healthcare providers, even across different institutions, networks, and specializations. When patients transition between primary care providers, specialists, or hospital networks, EHRs facilitate instant data transfer without requiring patients to manually transport paper records or authorize sharing. This level of integration reduces the risk of duplicated tests, conflicting prescriptions, or inconsistent treatment plans, ultimately improving patient outcomes [10]. Particularly in multidisciplinary care, where multiple providers contribute to a patient's treatment plan, EHRs help prevent gaps in communication and ensure holistic, coordinated care.

Modern EHR systems come equipped with clinical decision-support tools designed to enhance diagnostic accuracy, treatment planning, and risk mitigation. Providers such as Epic Systems, Cerner (now Oracle Health), and Allscripts integrate these tools into their platforms, offering features such as automated alerts for drug interactions, reminders for preventive care, and evidence-based treatment recommendations. For example, Epic's Best Practice Advisory (BPA) system provides real-time alerts to physicians about potential medication contraindications, while Cerner's Millennium EHR includes clinical decision support features that assist with sepsis detection and early intervention. These tools help reduce medical errors and improve patient safety [11]. Additionally, patient engagement platforms within EHR systems, such as MyChart by Epic, send automated reminders for vaccinations, lab tests, and lifestyle modifications, improving adherence to medical recommendations and preventive care measures.

Beyond clinical care, EHRs contribute significantly to administrative efficiency. These systems streamline routine tasks such as scheduling, billing, and interdepartmental communication. By integrating automation, EHRs reduce manual data entry errors, minimize paperwork, and optimize resource allocation. Studies have shown that EHR adoption leads to improved appointment scheduling, better workload distribution among healthcare staff, and efficient management of medical supplies [2]. Furthermore, digital billing systems enhance accuracy in insurance claims processing, reducing administrative burdens and expediting reimbursements.

EHR systems extend their utility beyond individual patient care by supporting medical research and public health monitoring. Through controlled access to anonymized data, researchers can analyze trends, study disease patterns, and assess healthcare interventions without compromising patient privacy. Many EHR platforms include built-in data analysis and export tools, allowing for efficient epidemiological studies and policy assessments [12]. Public health agencies, such as the CDC and NIH, leverage real-time EHR data to track disease outbreaks, monitor vaccination rates, and evaluate the effectiveness of healthcare policies [13]. The ability to integrate large-scale health data with emerging technologies such as machine learning and blockchain further enhances predictive analytics and personalized medicine.

As EHR systems continue to evolve, it is critical to ensure that technological advancements do not compromise their foundational objectives. Cutting-edge technologies, such as artificial intelligence and blockchain, offer promising improvements in security and efficiency but may also introduce challenges related to system stability and data accessibility. Ensuring that these innovations enhance rather than hinder the core functions of EHRs remains a central focus in modern healthcare IT development.

2.2 Security features

The core features identified above represent significant advancements in healthcare, but the digitization of healthcare records also introduces risks. As healthcare networks increasingly rely on digital systems, protecting sensitive EHR information becomes paramount. Storing patient data in digital formats exposes healthcare providers to cyber threats that were not previously a concern with paper-based records. However, modern EHR systems employ robust security features to mitigate these risks and safeguard patient information. Below are some of the most critical security measures implemented in contemporary EHR systems.

Encryption is a fundamental cybersecurity measure that ensures patient data remains secure. EHR systems encrypt data into unreadable formats, accessible only through proprietary decryption tools. This prevents unauthorized entities from exploiting data breaches. Advanced encryption standards, such as AES-256 (Advanced Encryption Standard with 256-bit keys), protect data both at rest (stored on secure hospital servers) and in transit (when shared with external entities such as specialist clinics, government organizations, or public health researchers) [14]. Only authorized personnel can access decryption tools, and data exported from hospital servers is encrypted with unique keys separate from those used for stored data. This layered encryption approach ensures that even if external data is decrypted, hospital-stored data remains secure.

Additionally, some EHR systems employ *homomorphic encryption*, allowing data to be analyzed while still encrypted. This minimizes the need for decryption, ensuring patient health information remains protected and is only accessed in raw form when explicitly necessary (**Figure 1**) [14].

Effective access control mechanisms are essential for limiting data access to authorized personnel. Most EHR systems implement Role-Based Access Control (RBAC), which ensures that users can only access data relevant to their role. For instance, physicians can access patient records only when assigned to a case or in emergency situations,

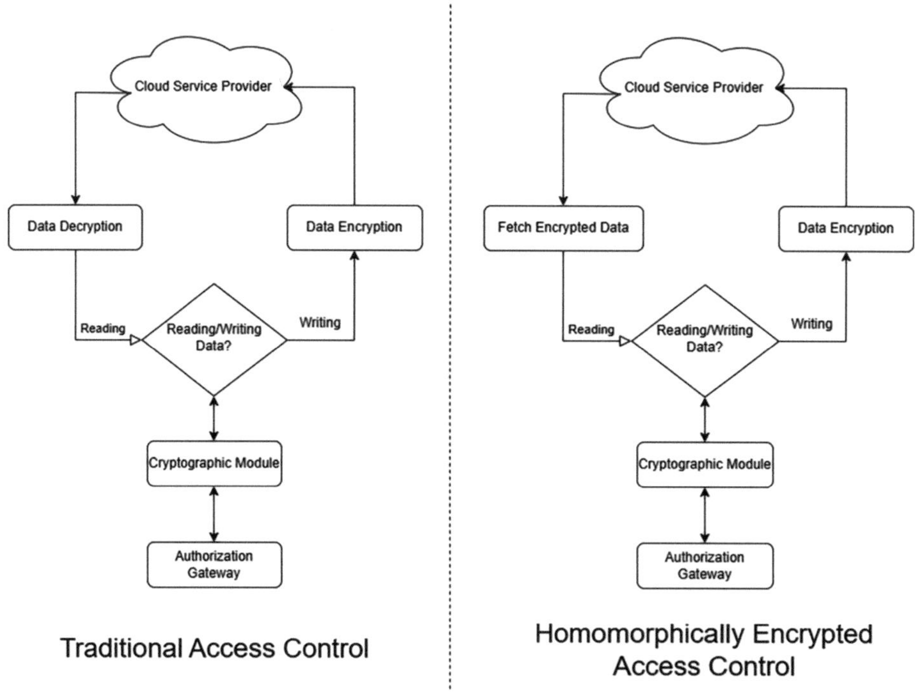

Figure 1.
Encryption in digital EHR systems.

while administrative staff may only have access to billing or demographic information. This minimizes the risk of insider threats and unauthorized data exposure [15].

MFA enhances security by requiring multiple verification methods before granting access. Typically, authentication involves a combination of a password, a one-time code sent to a registered device or email, and biometric verification (e.g., fingerprint or facial recognition).

Biometric authentication is particularly effective, as it provides a nearly foolproof identity verification method that is difficult to bypass. EHR systems increasingly integrate biometric authentication to add an extra layer of protection against unauthorized access [1].

Beyond preventing unauthorized access, it is critical to detect and block attempted cyberattacks. Intrusion Detection and Prevention Systems (IDPS) continuously monitor network traffic and system activity for suspicious behavior, such as multiple failed login attempts, unauthorized access requests, and unusual user activity.

When a potential attack is detected, IDPS takes proactive measures, such as blocking access from suspicious devices or accounts. These systems play a crucial role in mitigating threats such as malware, phishing attacks, ransomware, and brute-force login attempts [16].

If a cyberattack does occur, EHR systems need tools to assess the breach and prevent future incidents. Audit logs track all user interactions with the system, recording when and how data is accessed, modified, or shared. These logs help identify compromised accounts, determine the extent of a breach, and evaluate system vulnerabilities.

Audit logs also support ongoing cybersecurity improvements by pinpointing weaknesses in security measures. They provide valuable insights into system performance, helping administrators refine security protocols while maintaining system efficiency [17].

To safeguard against data loss due to cyberattacks, system failures, or user errors, EHR systems implement regular data backups stored in secure, offline locations. These backups ensure that data can be restored after a ransomware attack (or accidental deletion); healthcare operations continue with minimal disruption; and patient records remain intact even in the event of catastrophic system failures. EHR recovery mechanisms are designed for rapid data restoration, ensuring healthcare providers can access critical information without significant delays. Redundant data storage across multiple locations further enhances reliability and security [18].

Cybersecurity remains a top priority in EHR system development. The combination of encryption, access control, authentication, intrusion detection, audit logging, and data backups ensures that patient data remains secure while enabling seamless healthcare operations. As cyber threats continue to evolve, these security measures will be continuously refined to provide even stronger protection for sensitive health information.

3. Problems facing current EHR systems

While Electronic Health Record (EHR) systems have revolutionized healthcare by improving accessibility, efficiency, and security, they remain vulnerable to evolving cybersecurity threats. As digital security threats grow in complexity, healthcare institutions must continuously refine their cybersecurity strategies to protect sensitive patient data. This is particularly critical given the widespread and rapid adoption of digital EHR systems.

Ransomware is one of the fastest growing cybersecurity threats targeting EHR systems. In these attacks, malicious software encrypts patient data, preventing access until a ransom is paid. The healthcare sector is especially susceptible due to the urgent need for uninterrupted access to patient records. While data backups and recovery plans help mitigate these threats, they are not foolproof. Data loss still occurs if new information is added after the most recent backup, and recovery efforts take time, often causing operational disruptions [19].

Beyond operational risks, ransomware attacks impose severe financial burdens on healthcare institutions. The cost of healthcare data breaches is among the highest across industries, making hospitals and clinics prime targets. Cybercriminals exploit this by demanding higher ransoms, knowing that healthcare providers cannot afford prolonged downtime [19, 20]. This financial incentive explains why ransomware has become one of the most prevalent forms of cyberattacks in healthcare.

Phishing attacks are another major cybersecurity challenge for EHR systems. These attacks exploit human error by deceiving healthcare employees into divulging sensitive login credentials or downloading malware. Cybercriminals may use fraudulent emails, phone calls, or fake websites to trick individuals into providing access to EHR networks. Despite sophisticated cybersecurity measures, the human element remains a persistent vulnerability [21].

This risk is heightened by the rapid shift from paper-based records to digital systems. Many healthcare professionals, particularly those accustomed to traditional record-keeping, may not be fully trained in recognizing phishing attempts. This knowledge gap increases the likelihood of successful cyberattacks [21, 22]. Regular staff training and awareness campaigns are crucial to reducing this vulnerability.

Encryption is a key cybersecurity measure used in EHR systems to protect patient data. However, vulnerabilities still exist—particularly concerning "data at rest," or data that is stored but not actively being transmitted. Some healthcare systems do not encrypt stored data, assuming that restricted physical access to servers is sufficient protection. This assumption is flawed. If attackers gain physical access to a data center or exploit a system-wide failure, unencrypted patient records become easily accessible [21, 22]. To address this issue, modern EHR systems must implement full-disk encryption and other advanced security measures. Institutions should also adopt zero-trust security models, ensuring that even internal users must authenticate access at multiple levels [20].

Many healthcare organizations continue to rely on legacy systems—older software or hardware platforms that lack modern security features. These outdated systems may not support multi-factor authentication, advanced encryption, or real-time monitoring, making them easy targets for cybercriminals. Additionally, older operating systems often lack critical security updates, leaving vulnerabilities exposed [20].

A major obstacle to upgrading legacy systems is cost. Many hospitals operate on tight budgets and hesitate to overhaul IT infrastructure unless a system is completely nonfunctional. However, a "if it works, don't fix it" approach is dangerous in cybersecurity. By the time an unpatched system is exploited, it is often too late to prevent data theft or system-wide failure [20, 22]. Regular security audits, patch management, and phased modernization of IT infrastructure are essential to reducing risks associated with legacy EHR systems.

4. The future of EHR systems

The evolution of Electronic Health Record (EHR) systems has been instrumental in modernizing healthcare delivery, enhancing the accessibility, accuracy, and management of patient data. However, despite their numerous advantages, current EHR systems still face significant challenges, including security vulnerabilities, inefficiencies in usability, interoperability issues, and scalability concerns. Addressing these shortcomings is crucial for the next generation of EHR systems. Future advancements will integrate emerging technologies and cutting-edge innovations to create more effective, secure, and user-friendly systems.

One of the most promising developments in EHR technology is the integration of artificial intelligence (AI)/machine learning (ML). AI-driven security tools can analyze vast amounts of data in real time, detecting anomalies and patterns indicative of cyber threats. This predictive capability allows EHR systems to identify and prevent cyberattacks such as phishing and ransomware before they cause harm [23]. AI can automatically restrict access to suspicious devices or user activity, thereby enhancing security with minimal human oversight.

Beyond security, AI can significantly improve data integrity by identifying and flagging errors caused by human input, such as incorrect patient information. ML algorithms can detect inconsistencies and prompt human review, reducing the risk of misdiagnoses or inaccurate prescriptions [24]. Furthermore, AI can be leveraged for predictive analytics, identifying trends in patient data that may signal early signs of disease or gaps in care, ultimately improving patient outcomes [25].

A specialized subset of AI, natural language processing (NLP), is revolutionizing how EHR systems handle data entry and communication. NLP-powered transcription tools can convert physician-patient interactions into structured documentation, reducing the burden of manual data entry and minimizing errors [26]. This automation allows healthcare providers to focus more on patient care rather than administrative tasks.

Additionally, NLP can break down language barriers in healthcare by enabling real-time translation of medical records. Patients receiving care in international settings or from providers who speak a different language can benefit from seamless communication. NLP-driven chatbots and digital assistants can also aid patients in navigating healthcare portals, scheduling appointments, or retrieving medical information, particularly benefiting elderly or non-English-speaking patients (**Figure 2**) [27].

Blockchain technology presents a transformative opportunity for EHR systems by ensuring data security, integrity, and interoperability. By utilizing a decentralized ledger, blockchain records all access, edits, and additions to patient records in an immutable format, making unauthorized alterations nearly impossible [28]. This transparency helps prevent fraud and unauthorized data modifications, ensuring the reliability of medical records.

Furthermore, blockchain facilitates seamless data exchange between different healthcare systems. With a standardized and decentralized approach to data storage, hospital networks can securely share patient records while maintaining strict privacy controls. This eliminates inconsistencies in patient information across different healthcare providers and enhances coordination in patient care [4].

Interoperability remains a persistent challenge in current EHR systems. Fast Healthcare Interoperability Resources (FHIR) is an emerging standard that aims to

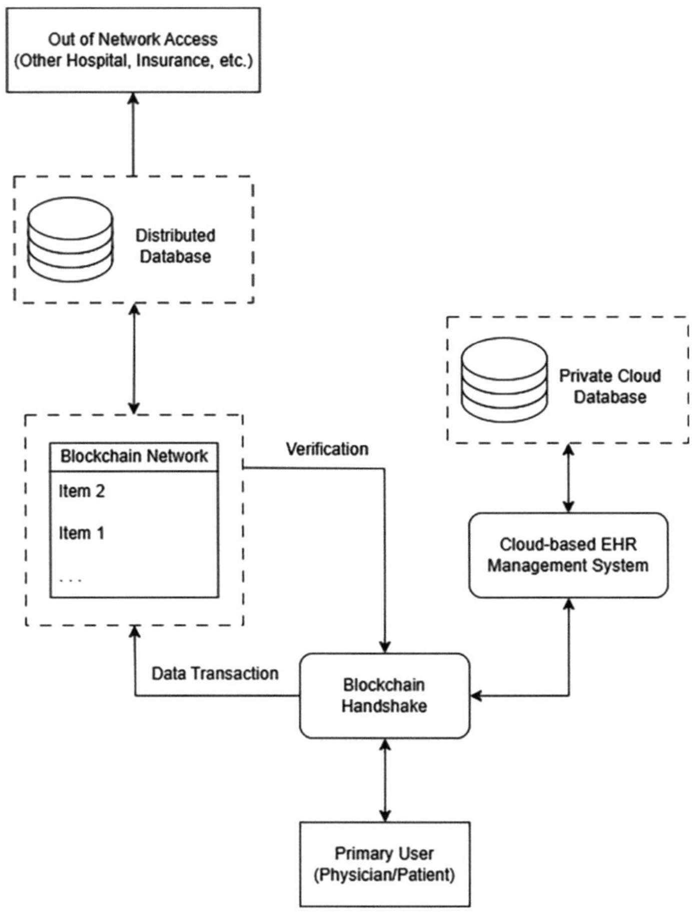

Figure 2.
Blockchain implementation in EHR systems.

improve the way health information is exchanged between providers, payers, and patients [29]. FHIR provides a flexible framework for storing and sharing EHR data in a structured, easily accessible format.

Additionally, many EHR vendors, including Epic and Cerner, are developing Application Programming Interfaces (APIs) to enable seamless data exchange between different systems. APIs allow external applications to interact securely with EHR platforms, facilitating automated data transfers, predictive analytics, and third-party integrations. While concerns about data security exist, APIs use authentication mechanisms, such as API keys, to restrict access to authorized entities, ensuring safe and efficient interoperability [30].

With the rapid expansion of healthcare data—including genomic information, biometric readings, and patient-generated health data—scalability is a growing concern. Cloud computing offers a solution by enabling hospital networks to store and manage vast amounts of patient data without requiring extensive on-premises infrastructure [31].

Major cloud providers such as Amazon Web Services (AWS), Microsoft Azure, and Google Cloud offer healthcare-specific solutions that provide on-demand scalability, automated data backups, and enhanced security compliance [32]. Cloud-based EHR systems reduce downtime caused by hardware failures and improve disaster recovery capabilities, ensuring uninterrupted access to critical medical information. These platforms also incorporate advanced encryption and security protocols, addressing many concerns related to data breaches and unauthorized access.

The future of EHR systems lies in leveraging AI, blockchain, interoperability standards like FHIR, and cloud computing to address current challenges while optimizing healthcare delivery. AI-powered tools will enhance security, data integrity, and predictive analytics, while NLP-driven automation will streamline data entry and communication. Blockchain technology will bolster security and interoperability, and FHIR standards, coupled with APIs, will enable seamless data exchange. Lastly, cloud computing will ensure that EHR systems remain scalable, reliable, and capable of handling the ever-growing volume of healthcare data. These advancements will drive the next generation of EHR systems toward greater efficiency, security, and accessibility, ultimately improving patient care and outcomes.

5. Discussion and conclusion

The security and privacy of Electronic Health Record (EHR) systems are critical to the future of healthcare. As the digitization of patient information continues to expand, safeguarding this sensitive data against evolving cyber threats has become a paramount concern. The implementation of encryption, access control mechanisms, and multi-factor authentication has significantly enhanced protection, yet these methods are not without challenges, such as computational overhead, integration complexities, and the need for continuous updates to stay ahead of emerging attack vectors.

Innovative technologies such as blockchain and artificial intelligence (AI) offer promising solutions to address some of these shortcomings. Blockchain, with its decentralized and immutable structure, enhances data security by preventing unauthorized modifications and providing transparent audit trails. AI-driven security tools, particularly those leveraging machine learning (ML), can analyze vast amounts of data in real time to detect anomalies, predict cyber threats, and enhance system efficiency. Moreover, AI-based natural language processing (NLP) can streamline clinical workflows by enabling automated transcription, real-time translation, and intelligent data validation. However, while these technologies present significant advantages, they also introduce new challenges related to scalability, adoption costs, and resource demands, particularly for larger healthcare systems.

Beyond security, the future of EHR systems will be defined by improved interoperability, enhanced usability, and expanded capabilities that leverage predictive analytics. As healthcare moves toward a more connected ecosystem, Fast Healthcare Interoperability Resources (FHIR) and API-driven integrations will play a crucial role in breaking down data silos, ensuring seamless communication between different EHR platforms, and enabling real-time data sharing between healthcare providers. This will improve care coordination, reduce redundant procedures, and enhance patient outcomes. Cloud computing is also expected to play a key role in the future of EHR systems, providing scalable storage solutions, reducing reliance on on-premises infrastructure, and improving system reliability while minimizing downtime.

Another major shift in the future of EHR systems is the incorporation of patient-centric models. Currently, EHRs are primarily designed for healthcare providers, but future iterations will increasingly prioritize patient engagement. Features such as decentralized health data ownership, patient-controlled access permissions, and AI-driven health insights will empower individuals to take a more active role in managing their health. Wearable devices and remote monitoring technologies will further integrate with EHR systems, providing real-time health data that allows for more personalized and proactive care.

While the future holds immense potential, challenges remain. Privacy concerns, regulatory compliance, and ethical considerations must be addressed as EHR systems continue to evolve. Cyber threats, including ransomware and data breaches, will require continuous innovation in cybersecurity measures, including adaptive AI-driven security frameworks and more robust encryption methodologies. Additionally, training and educating healthcare professionals on cybersecurity best practices will remain a critical component in mitigating human-related vulnerabilities such as phishing and social engineering attacks.

Ultimately, the future of EHR systems will be shaped by a combination of technological advancements, regulatory frameworks, and evolving healthcare needs. By embracing AI, blockchain, interoperability standards, and cloud computing, the next generation of EHR systems can offer a more secure, efficient, and patient-centric approach to healthcare. However, these advancements must be implemented with careful consideration of ethical, legal, and practical constraints to ensure that they truly enhance, rather than complicate, the healthcare experience. The ongoing evolution of EHR systems will not only improve data security and accessibility but will also transform the way healthcare is delivered, making it more predictive, personalized, and efficient.

Conflict of interest

The authors declare no conflict of interest.

Author details

Ritviksiddha Penchala[1] and Venkata Krishna Parimala[2]*

1 University of California San Diego, La Jolla, CA, United States of America

2 Department of Computer Science, Sri Padmavati Mahila Visvavidyalayam, Tirupati, India

*Address all correspondence to: pvk@spmvv.ac.in

References

[1] Bhargav-Spantzel A, Squicciarini AC, Modi S, Young M, Bertino E, Elliott SJ. Privacy preserving multi-factor authentication with biometrics. Journal of Computer Security. 2007;**15**(5):529-560. DOI: 10.3233/JCS-2007-15503

[2] Kruse CS, Frederick B, Jacobson T, Monticone DK. Cybersecurity in healthcare: A systematic review of modern threats and trends. Technology and Health Care. 2017;**25**(1):1-10. DOI: 10.3233/THC-161263. PMID: 27689562

[3] Keshta I, Odeh A. Security and privacy of electronic health records: Concerns and challenges. Egyptian Informatics Journal. 2021;**22**(2):177-183. DOI: 10.1016/j.eij.2020.07.003

[4] Kuo TT, Kim HE, Ohno-Machado L. Blockchain distributed ledger technologies for biomedical and health care applications. Journal of the American Medical Informatics Association. 1 Nov 2017;**24**(6):1211-1220. DOI: 10.1093/jamia/ocx068

[5] Zuo Z, Watson M, Budgen D, Hall R, Kennelly C, Al Moubayed N. Data anonymization for pervasive health care: Systematic literature mapping study. JMIR Medical Informatics. 2021;**9**(10):e29871. DOI: 10.2196/29871

[6] Scarfone K et al. Guide to intrusion detection and prevention systems (IDPS). NIST Special Publication. 2007:800-894. DOI: 10.6028/NIST.SP.800-94

[7] Legacy systems in healthcare impact growth, data security [Internet]. 2023. Available from: https://thecyberexpress.com/legacy-systems-in-healthcare-hinder-growth/

[8] Mohandas R, Sundareswaran V, Kirubasri G, Thusnavis BM, Udayakumar R. Federated learning with homomorphic encryption for ensuring privacy in medical data. Indian Journal of Information Sources and Services. 2024;**14**(2):17-23. DOI: 10.51983/ijiss-2024.14.2.03

[9] de Carvalho Junior MA, Bandiera-Paiva P. Health information system role-based access control: Current security trends and challenges. Journal of Healthcare Engineering. 2018;**2018**:6510249. DOI: 10.1155/2018/6510249. PMID: 29670743; PMCID: PMC5836325

[10] Basil NN, Ambe S, Ekhator C, Fonkem E. Health records database and inherent security concerns: A review of the literature. Cureus. 11 Oct 2022;**14**(10):e30168. DOI: 10.7759/cureus.30168. PMID: 36397924; PMCID: PMC9647912

[11] Pool J, Akhlaghpour S, Fatehi F, Burton-Jones A. A systematic analysis of failures in protecting personal health data: A scoping review. International Journal of Information Management. 2024;**74**:102719. DOI: 10.1016/j.ijinfomgt.2023.102719

[12] He Y, Aliyu A, Evans M, Luo C. Health care cybersecurity challenges and solutions under the climate of COVID-19: Scoping review. Journal of Medical Internet Research. 20 Apr 2021;**23**(4):e21747. DOI: 10.2196/21747

[13] Shenoy A, Appel JM. Safeguarding confidentiality in electronic health records. Cambridge Quarterly of Healthcare Ethics. Apr 2017;**26**(2):337-341. DOI: 10.1017/S0963180116000931. PMID: 28361730

[14] Jiang F, Jiang Y, Zhi H, Dong Y, Li H, Ma S, et al. Artificial intelligence in healthcare: Past, present, and future. Stroke and Vascular Neurology. 2017;**2**(4):230-243

[15] Rajkomar A, Dean J, Kohane I. Machine learning in medicine. New England Journal of Medicine. 2019;**380**(14):1347-1358

[16] Wu W, Holkeboer KJ, Kolawole TO, Carbone L, Mahmoudi E. Natural language processing to identify social determinants of health in Alzheimer's disease and related dementia from electronic health records. Health Services Research. 2023;**58**(6):1292-1302. DOI: 10.1111/1475-6773.14210

[17] Roberts K, Kilicoglu H, Fiszman M, Demner-Fushman D. Automatically classifying question types for consumer health questions. AMIA Annual Symposium Proceedings. 14 Nov 2014;**2014**:1018-1027. PMID: 25954411; PMCID: PMC4420005

[18] Engelhardt MA. Hitching healthcare to the Blockchain: A regulatory and legal framework. The Journal of Health Care Law and Policy. 2017;**20**(1):1-40

[19] Mandel JC, Kreda DA, Mandl KD, Kohane IS, Ramoni RB. SMART on FHIR: A standards-based, interoperable apps platform for electronic health records. Journal of the American Medical Informatics Association. 2016;**23**(5):899-908

[20] Reddy S. Generative AI in healthcare: An implementation science informed translational path on application, integration, and governance. Implementation Science. 2024;**19**(1):27. DOI: 10.1186/s13012-024-01357-9

[21] Griebel L, Prokosch HU, Kutschek B, Toddenroth D, Christoph J, Leb I, et al. A scoping review of cloud computing in healthcare. BMC Medical Informatics and Decision Making. 2015;**15**(1):17

[22] Dixon BE, Grannis SJ, Lenert LA. Leveraging data visualization and a Statewide health information exchange to support COVID-19 surveillance and response: Application of public health informatics. Journal of the American Medical Informatics Association. 2019;**26**(5):450-455

[23] Melton GB, McDonald CJ, Tang PC, Hripcsak G. Electronic health records. In: Shortliffe EH, Cimino JJ, editors. Biomedical Informatics. Cham: Springer; 2021. pp. 467-509. DOI: 10.1007/978-3-030-58721-5_14

[24] National Institute of Standards and Technology (NIST). Health IT and EHR Interoperability [Internet]. 2021. Available from: https://www.nist.gov/health-it

[25] Centers for Disease Control and Prevention. PLACES [Internet]. 2024. Available from: https://www.cdc.gov/places

[26] Alanazi AT. Clinicians' perspectives on healthcare cybersecurity and cyber threats. Cureus. 14 Oct 2023;**15**(10):e47026. DOI: 10.7759/cureus.47026. PMID: 37965389; PMCID: PMC10642560

[27] HealthIT.gov. Guide to privacy and security of electronic health information [Internet]. 2024. Available from: https://www.healthit.gov

[28] Mahendiran A, Appusamy R, Karthik S. Intrusion detection and prevention system: Technologies and challenges. International Journal of Applied Engineering Research. 2015;**10**:1-12

[29] O'Malley AS, Grossman JM, Cohen GR, Kemper NM, Pham HH. Are electronic medical records helpful for care coordination? Experiences of physician practices. Journal of General Internal Medicine. Mar 2010;**25**(3):177-185. DOI: 10.1007/s11606-009-1195-2. PMID: 20033621; PMCID: PMC2839331

[30] Bates DW, Kuperman GJ, Wang S, Gandhi T, Kittler A, Volk L, et al. Ten commandments for effective clinical decision support: Making the practice of evidence-based medicine a reality. Journal of the American Medical Informatics Association. Nov-Dec 2003;**10**(6):523-530. DOI: 10.1197/jamia.M1370. PMID: 12925543; PMCID: PMC264429

[31] Hillestad R, Bigelow J, Bower A, Girosi F, Meili R, Scoville R, et al. Can electronic medical record systems transform health care? Potential health benefits, savings, and costs. Health Affairs (Millwood). Sep-Oct 2005;**24**(5):1103-1117. DOI: 10.1377/hlthaff.24.5.1103. PMID: 16162551

[32] Friedman CP, Wong AK, Blumenthal D. Achieving a nationwide learning health system. Science Translational Medicine. 10 Nov 2010;**2**(57):57cm29. DOI: 10.1126/scitranslmed.3001456. PMID: 21068440

Section 2

Advances in Mobile and AI-Driven Health Technologies

Chapter 4

Emerging Mobile Health Systems and Services

Theodoros Solomou, Christos N. Schizas and Constantinos S. Pattichis

Abstract

The advancement and integration of Electronic Health Record (EHR) systems into mobile solutions have transformed healthcare by offering a wide array of new services. This chapter explores the pivotal role of mobile-based solutions in national healthcare, emphasizing their accessibility, portability, and ability to integrate smart devices and sensors to enhance patient care. We investigate various country-specific implementations of mobile EHR solutions, analyzing technical architectures, interoperability standards, and the challenges these systems face. Additionally, our study examines how these solutions effectively address the limitations of traditional healthcare settings by providing seamless, secure access to health data, improving healthcare delivery efficiency, and ensuring patient-centered care through innovative technologies, including AI-driven solutions and coaching systems. By evaluating these aspects, we aim to shed light on the effectiveness, user-friendliness, and potential impact of mobile EHR systems on healthcare experiences. These efforts are supported now by the recently voted European Health Data Space regulation. Ultimately, we advocate for a paradigm shift toward more integrated, patient-oriented healthcare solutions that benefit citizens.

Keywords: mHealth, eHealth, digital health, national implementation, accessibility, interoperability

1. Introduction

In recent years, many countries have been investing in the digitalization of their healthcare sector. Their main focus is the implementation of EHR systems, which often take priority due to their core significance and functionality. EHR systems enable the identification of citizens, access to their medical history, and updates to records for future care. In these developments, a key component is the integration of the services in a mobile application. Smartphones are the most accessible devices for citizens, utilized daily, and can offer additional features enhancing the use of EHR, such as teleconsultation and monitoring. mHealth can be defined as the emerging mobile communications and network technologies for healthcare systems [1].

Patient access to medical records is a pressing global concern. The objective is to empower individuals to manage their own health data. This effort aligns with the recent developments in the European Health Data Space Regulations (EHDS), which were approved in April 2024 [2]. Notably, Article 8 of the EHDS Regulation [3] mandates that "Natural persons shall have the right to access their personal electronic health data processed in the context of primary use of electronic health data, immediately, free of charge, and in an easily readable, consolidated, and accessible form." This regulation ensures that, in the near future, every European citizen will have direct access to their medical data, with other continents likely to adopt similar measures. The EHDS also requires the provision of mechanisms that enable patients to share their data, reflecting the rapid evolution of the healthcare landscape toward enhanced patient access and enriched service offerings. This chapter is motivated by these significant changes that prioritize patient empowerment and the expansion of health services beyond traditional offerings.

Our objective is to explore EHR driven mobile solutions and how they are shaping to provide new possibilities. We will delve into the various emerging technologies that can be utilized for EHR, understand what the benefits are of using a mobile-based solution, and define the main components of a national mobile app for EHR. Additionally, we will take a closer look at country-specific implementations. By examining these solutions, we can understand the range of features they offer beyond basic access. This comprehensive analysis will provide insight into the effectiveness and user friendliness of the available EHR solutions and their potential impact on the healthcare experience for citizens.

In the following section, an introduction is made to mobile EHR services. In Section 3, we explain the available technologies that can be combined with mHealth applications. In sections 4 and 5, we discuss the technical requirements and implementation strategies for developing an mHealth solution. In Section 6, we take a closer look to national country-specific solutions and Section 7 describes the challenges of such systems alongside future trends. Finally, in Section 8, we provide concluding remarks.

2. Overview of mHealth services

The doctor of the future will be empowered by mobile health systems and integrated electronic health records (EHR), transforming the way healthcare is delivered. Equipped with real-time access to comprehensive patient data, including biometric information from wearable devices and detailed medical histories from EHRs, doctors will be able to make more accurate diagnoses and provide personalized care. Mobile health systems will allow healthcare providers to seamlessly conduct teleconsultations, prescribe medications through ePrescriptions, and monitor patient progress remotely, all while reducing the administrative burdens that currently hinder their efficiency. This shift will enable doctors to focus more on patient interaction and high-quality care, enhancing outcomes and making healthcare more accessible and efficient for everyone.

In recent years, the technological landscape has rapidly evolved with advancements in AI, blockchain, and 5G, opening new opportunities to enhance citizens' quality of life. AI, particularly in prediction models and generative systems, has become accessible, even allowing voice interactions *via* smartphones [4]. In healthcare, AI-powered clinical support systems offer valuable second opinions, aiding professionals in diagnosis. Studies have shown promising results in implementing these technologies in healthcare settings, demonstrating their potential to significantly improve patient care and decision-making processes [5].

As a result, we are moving away from the traditional mobile services where patients passively viewed their medical data, toward a new model of patient engagement. In this new approach, the patient takes an active role in managing their health. Emerging services such as digital wallets, teleconsultations, and health monitoring technologies empower patients to actively participate and take control of their healthcare (see **Figure 1**).

The cornerstone of a traditional national EHR system is its ability to provide comprehensive access to a citizen's medical history, including essential information such as allergies, medications, and past medical procedures (see **Figure 1**). This enables healthcare professionals to quickly assess a patient's health data, reducing the risk of missing vital details often overlooked in verbal communication. These systems are typically built on models like the International Patient Summary [6], which support key cross-border healthcare scenarios.

In addition, MyHealth@EU is a European digital health network that facilitates the secure exchange of health data across EU member states, with all member countries working diligently toward its full implementation [7]. Beyond the basic patient summary, the system also includes laboratory and imaging results, alerts, and episodes of care. Alerts notify users of critical health indicators, while episodes of care document treatments received during healthcare visits, providing a clear, continuous overview of a patient's health journey.

As healthcare and mHealth systems evolve, new services are being integrated into EHR systems to enhance patient engagement and streamline care (see **Figure 1**). These services, many of which are accessible *via* mobile platforms, empower patients to manage their health more proactively while ensuring seamless communication with

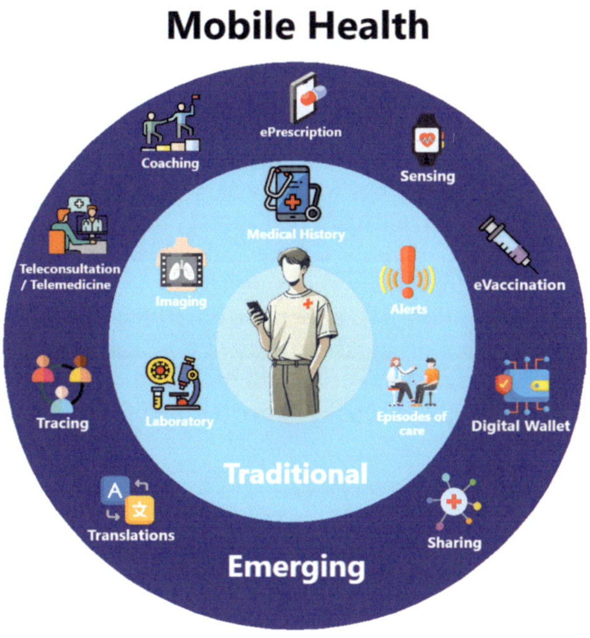

Figure 1.
Emerging services introduce new opportunities where patients take an active role in managing their health and have greater control, as opposed to traditional healthcare services, which primarily focus on managing healthcare documents.

healthcare providers. For example, teleconsultation is becoming increasingly popular, especially on mobile platforms due to their accessibility and user-friendly interfaces. For routine checkups and short meetings, teleconsultation can provide a convenient and quick solution, saving time for both the patient and doctor. Additionally, this service can be linked with calendar systems and management of online appointments, enhancing communication between patients and healthcare providers [8].

Another important emerging mobile health service is ePrescription. In Europe, it is a key component of the digital transformation of healthcare, allowing the electronic transfer of prescriptions from healthcare providers to pharmacies across EU member states. This service is part of the broader MyHealth@EU initiative [7], as mentioned previously. The implementation of ePrescriptions is supported by the European eHealth Network and the EU's Cross-Border Healthcare Directive (2011/24/EU) [7], which ensures that citizens can use healthcare services across EU borders.

The secure sharing of medical data through smartphones will play a crucial role in the future, enabling patients to maintain a single, accessible repository of their health records, which integrates with various health-sensing devices like smartwatches. Additionally, during the COVID-19 pandemic, the introduction of eVaccination and contact tracing services underscored the importance of mobile health applications [9]. Vaccination certificates are securely stored and easily verified *via* QR codes, while contact tracing used GPS data to track disease spread efficiently, providing crucial data for public health management. Further enhancing the EHR system, health coaching through mobile platforms can offer personalized guidance and wellness tracking, helping individuals achieve healthier lifestyles based on real-time data from their daily activities and medical history. Moreover, integration with the national Digital Wallet under EHDS regulations [10] ensures that all health data transactions meet stringent security and privacy standards, making mobile health management both convenient and compliant.

In **Table 1**, we present the advantages of using a mobile EHR solution. The shift to mobile EHR systems brings significant benefits over traditional models primarily due to their ubiquitous accessibility. People typically carry their smartphones everywhere, leveraging their portability unlike desktop computers that require stationary use, often confined within hospital environments. Mobile access not only enables viewing of medical information from any location with internet access but also supports offline data storage for emergencies, allowing paramedics immediate access to vital information through devices like smartwatches. Additionally, the integrated sensors on smartphones, such as microphones for voice recognition [11], enrich the functionality of EHRs by facilitating easier teleconsultations and improving the quality of medical documentation.

Mobile EHRs shift data management to patients, empowering them to actively participate in their health management by contributing biometric data through smart

Advantages of using a mobile EHR solutions		
Accessibility	**Portability**	**Offline use**
Access to plethora of sensors	Video teleconsultation	Improved communications
Simpler UI	Ease of use	Interoperability
Reduced costs	Biometric authentication	Notifications

Table 1.
Advantages of mobile EHR solutions.

sensing, which we will cover in Section 3.2. This model, however, introduces significant security challenges that require robust measures to protect sensitive patient information. Additionally, the interactive capabilities of smartphones, including real-time notifications and messaging, facilitate more dynamic and efficient communication between patients and healthcare providers, contrasting with the more static and often interruptive communication methods of traditional web portals.

Furthermore, mobile platforms revolutionize how health data are managed and communicated. Unlike traditional EHR systems where data are localized within hospital servers, mobile EHRs often utilize cloud-based solutions, which we discuss at Section 3.3, that aggregate data from multiple healthcare organizations into a centralized database accessible from a smartphone. This not only breaks down data silos and enhances interoperability but also reduces the infrastructure costs associated with traditional systems, allowing for further investments in service improvement.

3. Mobile health technologies

National mobile EHR systems rely on a diverse array of communication technologies to ensure the seamless transfer of health data across various platforms and devices. Technologies such as 5G, Wi-Fi, Bluetooth, and satellite links each play a unique role in enhancing connectivity, with ongoing advancements increasing bandwidth and reducing latency. While these networks enable more efficient data sharing, they also introduce security challenges, especially when dealing with sensitive health information. As we delve into the evolving landscape of communication technologies, it is essential to consider how they contribute to both the functionality and security of mobile EHR systems, as well as their integration with smart devices and cloud-based solutions that further optimize healthcare delivery and patient monitoring.

3.1 Communication links

In the domain of national mobile EHR systems, communication technologies form the critical infrastructure that facilitates the seamless exchange of health data across various platforms and devices. These technologies, including cellular networks (5G), Wi-Fi, Bluetooth, and satellite links, each fulfill a unique role. Annually, these communication links evolve, delivering increased bandwidth and faster speeds that significantly decrease latency. This advancement particularly enhances smartphone functionality, improving user experience with more stable connections and quicker response times.

However, the transmission of sensitive data through these networks raises security concerns. Health applications commonly use technologies such as SSL/TLS to protect data during transmission. Despite adherence to best practices, some communication channels may not be suitable for certain contexts. For example, the eIDAS 2.0 regulation [12] for the European Digital Wallet recommends user authentication *via* Bluetooth rather than cellular data. This preference is due to the physical presence required for Bluetooth, which adds a layer of security.

With 5G networks, mobile EHR systems can support more sophisticated and data-intensive services. For instance, real-time remote patient monitoring becomes more feasible and reliable [13]. Wearable devices and IoT sensors can continuously collect patient data and transmit it instantaneously to healthcare providers. A notable example of such applications is a pilot study that involved the remote monitoring of cardiac patients and the provision of timely interventions [14]. The study found that patients who participated in this approach experienced significantly better rehabilitation outcomes compared to those following traditional methods.

3.2 Smart devices, internet of things (IoT), and AI

The most prominent smart device is the smartphone. Monitoring mobile health apps allow users to track and monitor their health and wellness metrics. These apps can help users keep track of important health information such as blood pressure, heart rate, sleep patterns, and physical activity (see also **Table 2**). Monitoring mobile health apps provides users with a wealth of information and insights to their health status and can be used to detect potential problems. Additionally, their sensors are becoming an increasingly popular method for tracking and recording health data. These sensors are capable of gathering extensive information about an individual's physical activity, heart rate, sleep patterns, and other biosignals, offering a cost-effective method for health monitoring. In this response, we explore two studies [15, 16] that present various uses of mobile sensors as listed in **Table 2**.

Table 2 highlights the capability of smartphone cameras to monitor a wide range of biosignals. This is made possible by AI systems, often utilizing deep learning and machine learning techniques, that analyze captured images or videos to predict or identify health conditions. Beyond static analysis, these models can also compare images taken over time, enabling them to track the progression of a disease or physiological state. While developing such AI systems requires significant effort and resources, our key argument is that modern smartphones, equipped with advanced sensors, can gather an impressive amount of health-related data. This wealth of information opens the door to more accessible, mobile-driven health monitoring, reducing the need for specialized devices.

Beyond smartphones, we also have tablets, smart goggles, and smartwatches embedded with numerous sensors such as heart rate monitors, barometer, accelerometer, respiration. These devices are all part of the expanding internet of things (IoT) landscape [17], a network of interrelated devices that connect and exchange data with other IoT devices and the cloud [18]. Smart scales, connected inhalers, vital monitors,

Vital	Sensor	Method
Heart rate	Camera, flashlight	Photoplethysmography: Using a phone's flashlight and camera, light variations through a finger are analyzed to detect heart rate.
Respiration	Camera, microphone	Record breathing or detect chest movement
Skin condition	Camera	Image from camera can be used in a prediction model to detect a medical condition.
Eye condition	Camera	Image from camera can be used in a prediction model to detect a medical condition.
Ear condition	Microphone	Play sounds in specific frequencies and collect users' input
Cognitive functions	Motion sensors, GPS	Based on movement a user can be evaluated if he/she is disoriented or lost.
Daily activity	Motion sensors, GPS	Tracks steps and exercise activity

Table 2.
Smartphone sensors used for monitoring vital sign and how.

are integral for managing various health conditions and generate an enormous number of data to be analyzed. The data collected are crucial for monitoring patients remotely, providing timely interventions, and shaping personalized care plans.

AI for patient monitoring is revolutionizing healthcare by leveraging medical devices to continuously track vital signs and detect intra-subject variability. Such a concept provides the opportunity for the establishment of individual trends and behavioral patterns, hence giving the ability for the early detection of potential health problems well before critical stages are reached. By using advanced algorithms, these systems can offer personalized feedback, based on the evolution of the patient's health status, and coach behaviors that lead to healthier outcomes [19]. This is an essential part of a mobile EHR system, tracking the health of the patient and coaching him on how to improve [20].

AI-driven solutions have become integral to mobile health systems, offering personalized care and improving decision-making processes. AI technologies, including machine learning algorithms and predictive models, enable healthcare providers to derive insights from patient data, thus enhancing the accuracy of diagnostics and treatment planning [5]. Moreover, AI-powered clinical decision support systems provide valuable second opinions, helping healthcare professionals in diagnosing complex conditions and optimizing patient outcomes [5]. Finally, generative AI is able to explain its reasoning, that can lead to better communication with the patients to understand their disease or justify the prediction to the healthcare professional [21, 22].

3.3 Cloud computing

Cloud computing is becoming a revolutionary tool in mobile EHR systems, fostering innovation and increasing efficiency in health care delivery [23]. With cloud technology, mobile EHR can securely store an enormous amount of medical data, and users such as healthcare professionals and patients can access data anywhere and anytime. Cloud architectures allow for easy scalability and flexibility, meaning that you can have a cost-effective solution and always have available resources for processing when needed. This becomes even more important as mHealth apps grow to include full-medical record datasets including imaging studies, lab results, and real-time patient monitoring data.

Security is a big issue with cloud-based mobile EHR systems, due to the fact that they are dealing with personal health information. Data are protected using the latest encryption techniques, secure data transfer protocols, and strict access controls. Cloud providers also usually have very strong disaster recovery and data backup plans in place so that in the event of hardware failure, natural disaster, or cyber-attack the data are still available. This resilience is essential for maintaining continuous healthcare services and building trust in mobile health solutions.

The integration of cloud computing into mobile EHRs also opens up new avenues for using big data analytics and machine learning to improve healthcare services [24]. They can process millions of data points to find trends, predict diseases, or support doctors with their clinical decisions. Also, cloud-based mobile EHRs can facilitate remote patient monitoring and telehealth that is playing a growing role in chronic disease management and in the delivery of health care to the medically underserved. In general, cloud computing empowers mobile EHR systems to perform more effectively and expansively, but also leads to more individualized, efficient, and available health care solutions.

4. Technical architecture and standards

Adoption of interoperability standards such as Fast Healthcare Interoperability Resources (FHIR) and Clinical Document Architecture (CDA) is essential in order to facilitate smooth data exchange of medical data between systems. While ensuring security and privacy, these allow sharing in real time, hence making it possible for modern web technologies and protocols to communicate between them. Besides, incorporating these technologies into today's healthcare facilities through mobile apps is crucial because it leads to better patient care as well as availability of critical information for medical practitioners at their closest point of need.

4.1 Interoperability standards and protocols

There is a need for a strong mechanism to efficiently and reliably exchange the medical data used in health care. However, there were no such well-defined protocols until recent years and one of them is the FHIR. FHIR is an emerging standard from HL7 International, giving developers new possibilities to easily speed up the flow of essential data across systems without extra integration costs. This has tried to bring together the best of HL7 prior versions with modern web standards and, at the same time, implement both simplicity in the implementation processes and improved interoperability between different systems [25].

Access, administration, and employability of health information are made smoother by FHIR in mobile health apps. Patient records can be viewed and updated anytime by their owners using the application. Clinical Decision Support Systems running on a mobile device use patient detail to make suggestions that would help a doctor in making decisions concerning treatment of illness or prevention of diseases. In addition, FHIR has enabled linking with health tracking gadgets or wearables, making it possible for clinicians to get continuous streams of information about their patients' health [26]. Furthermore, these applications help with prescription adherence and follow-up hence enabling better medication management. Mobile research also benefits from FHIR since it brings about uniformity in data collection while emergency response applications benefit by delivering vital patient information to ambulance services at accident scenes.

Another interoperability standard is the CDA, also developed by HL7 [27]. This standard is currently used in cross-border infrastructures such as OpenNCP, a secure network of national contact points that exchange healthcare data using the CDA format. CDA documents have a specific structure in XML format, and there are tools available for the generation and validation of these documents. CDA serves as a precursor to FHIR, which is gradually replacing it due to its greater robustness and numerous improvements. However, CDA documents will continue to be used, particularly in legacy systems, as they are a proven standard that facilitates interoperability among different systems for healthcare document exchange.

Another important standard is the Master Value Catalog (MVC). The MVC is a comprehensive collection of terms used within certain parts of the Patient Summary and ePrescription documents. These terms are based on standardized code systems such as ICD-10, SNOMED CT, ATC Classification, EDQM Standard Terms, and UCUM [28]. The MVC provides a common identification for diseases and medications, enhancing communication between countries and minimizing errors due to miscommunication. Each country maintains its own MVC, mapping the codings of terms to the terminology used in its official language (**Table 3**).

Fast healthcare interoperability resources (FHIR)	Clinical document architecture (CDA)	Master value catalog (MVC)
Utilizes RESTful APIs	XML format with specific structure	Collection of terms
Versatile and can define your own resources from primitive types	Variety of tools for generation and validation of documents	Standardized code system
Defined data types and you can create complex ones	Interoperability among different systems	Common identification of medical terms
Modular components	Used in the OpenNCP network	Enhances communication

Table 3.
Key characteristics of interoperability standards.

The final standard we will examine is ISO 13972:2022 [29]. This protocol serves as the foundations in health informatics by guiding the development and implementation of clinical information models (CIMs). This standard emphasizes the importance of creating universally understandable and interoperable clinical content, ensuring that data not only move seamlessly across diverse healthcare platforms but also retain its clinical relevance and integrity. ISO 13972 aims to enhance the consistency and accuracy of electronic health records, thus supporting improved clinical decisions and better patient outcomes. Additionally, it advocates for the inclusion of a broad range of stakeholders in the design process, ensuring that these models are practical and applicable in real-world clinical settings.

Finally, it is important to highlight the Label2Enable project, an EU-funded initiative aimed at promoting a quality and reliability label for health apps based on the CEN-ISO/TS 82304-2 standard [30]. This standard provides a comprehensive assessment framework, consisting of 81 questions that evaluate mHealth apps in areas such as ease of use, security, robustness, and the safety of their services. Upon completing the assessment, a detailed report is generated, offering an evaluation of the app's overall quality. The goal of Label2Enable is to introduce this label across all mHealth apps in mobile app stores, ensuring transparency and trust for users.

4.2 Security and privacy considerations

A vital part of EHR systems is security. There are a lot of sensitive data stored in these systems and we must ensure that no one can have access to these data unless it is the appropriate user, or he granted consent to someone to view his health data. For this reason, guidelines must be followed in order to achieve the desired outcome.

With smartphones readily available, we can take advantage of their built-in features to enhance our security. Common options such as fingerprint and face recognition provide an extra layer of protection against malicious attacks and are easily implemented thanks to sophisticated libraries that offer seamless access to these technologies. Additionally, there are less commonly used methods available on smartphones, such as iris recognition, voice recognition, and palm or finger vein patterns, all of which contribute to a robust security framework.

In a national EHR system, it is usually connected with multiple independent systems. There are many ways to implement this and manage communication between the services [31], as shown in **Figure 2**. The "One-service-one-login" model provides a single account for accessing all connected services, ideal for simplicity and unified

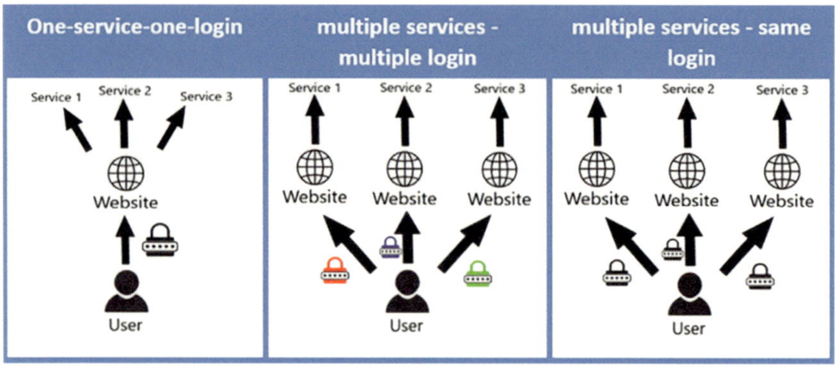

Figure 2.
The three approaches of connecting users to numerous services. The user could have a single account for all services, or have multiple accounts for each service or have one account that connects to each service with an authentication platform.

data management. However, this approach is less common due to the complexities involved in integrating different systems. The "multiple services-multiple logins" approach is more prevalent. It involves separate accounts for each service, aligning with the existing setups of individual organizations. Finally, the "multiple services-same login" strategy offers a balanced solution. It utilizes a national identification system to allow access to multiple services through a single login. This method simplifies user access without requiring a comprehensive overhaul of existing systems, making it a realistic and efficient option for enhancing interoperability among diverse healthcare providers.

Regarding security, the countries have different approaches—some safer than others. As we see in **Figure 3**, "One factor authentication" is when you have to just login with a username and a password only the citizen knows. "Multi-factor authentication" is similar to "one-factor" but when the user inserts the correct credentials, he must also verify his identity by providing a random number which is generated from the system and sent either through SMS, email, or certificate card with an embedded chip. This is obviously more secure since the user will be notified if a hacker successfully inserted his credentials, and a message is sent. "Multi-factor" authentication is the de facto solution which we often see in a mobile EHR system.

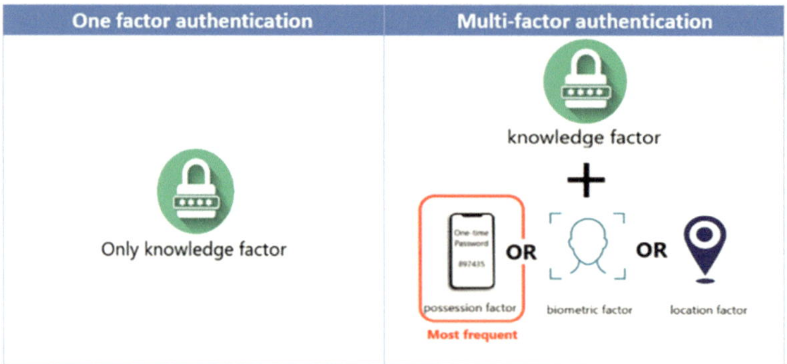

Figure 3.
One-factor authentication only requires a password, while multi-factor authentication requires additional authentication (possession, biometrics, or location).

Additionally for privacy, all data must be encrypted in the database. This is non-negotiable because in the case of a breach, even if they steal data, it will be hard to decrypt it and get the meaningful data [32]. Additionally, to preserve privacy of users, these systems must have data anonymization. There are specific attacks based on demographics or information that does not seem this important, which enable attackers to identify users and see what medical conditions they have. This is a violation of someone's privacy and with data anonymization you prevent the identification of users by grouping them in bigger groups [33].

4.3 Integration with existing healthcare infrastructure

Upgrading and integrating new protocols and standards into a national healthcare system, which maintains extensive medical data, can be challenging. Rather than starting from scratch, it is more practical to enhance existing systems by incorporating new components into the current infrastructure.

As we mentioned before, FHIR is an adaptive interoperability protocol we can integrate. It requires minimal adjustments to existing infrastructure, allowing medical information to be exchanged in a specified structure. **Figure 4** demonstrates how to facilitate this, as an intermediate layer between the existing healthcare infrastructure and the entities requesting data. This layer serves to retrieve data from national systems, convert it to the required format, incorporate additional information, and introduce new data management features.

Additionally in the figure we can see a mobile application that is compliant with Substitutable Medical Applications, Reusable Technologies on Fast Healthcare Interoperability Resources (SMART on FHIR) [34]. It is an open specification that allows developers to write apps effectively working within the systems of existing healthcare, in particular, from EHRs. This framework allows a Json, XML and SOAP-structured coding in the EHR system by third-party vendors, including software firms, which provide their own applications for health management, such as decision support, visualization of data, and patient engagement applications.

Additionally, RESTful mobile-specific APIs can be seamlessly integrated into the national infrastructure. These APIs are particularly suited for mobile networks due to their simplicity and stateless nature, which are advantageous given bandwidth

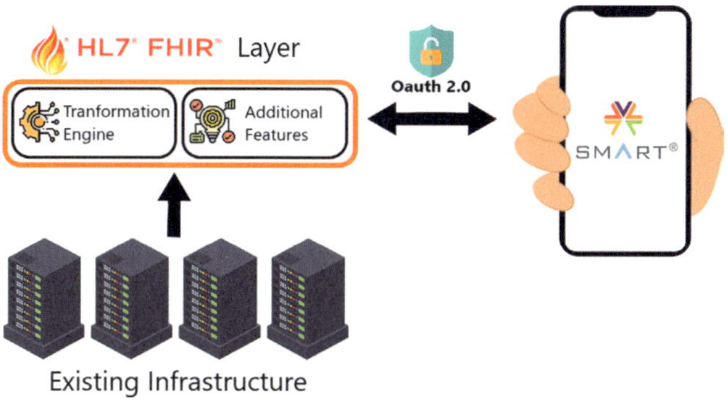

Figure 4.
System architecture of FHIR integration in mobile applications. An additional layer is introduced between the existing infrastructure and smartphones, converting data into FHIR resources and offering enhanced functionality. On the smartphone side, the system follows SMART on FHIR guidelines, communicating with the FHIR layer via a secure OAuth 2.0.

limitations and latency concerns. By offloading much of the processing from smartphones to government servers, these APIs enhance functionality without imposing heavy demands on mobile devices. Furthermore, implementing new APIs is scalable and adaptable but requires careful planning and development time. These components can be published in the intermediate level as shown in **Figure 4**.

In terms of security for a national mobile EHR system, leveraging OAuth 2.0 [35] through APIs enhances protection by enabling access to server resources without storing user credentials on mobile devices. Instead, access tokens are used, allowing healthcare professionals secure access to patient information without risking sensitive data exposure. Additionally, mobile devices equipped with biometric capabilities—such as fingerprint and facial recognition—can enhance security further by serving as components of two-factor authentication systems, providing a more robust security layer compared to traditional methods like one-time passwords.

5. Implementation strategies

The rapid growth of mHealth requires strong regulatory frameworks like the General Data Protection Regulation (GDPR) to protect personal data, while initiatives such as the EHDS focus on secure cross-border health data sharing. Developing a national mobile EHR system demands scalable, secure infrastructure, combining centralized servers with cloud solutions. Successful implementation relies on early engagement with key stakeholders, including healthcare professionals, citizens, and tech companies, to ensure that the systems developed are user-friendly, secure, and widely adopted.

5.1 Policy and regulation

The fast pace of technology development in mHealth brings special challenges for regulators around the world. However, initiatives such the GDPR [36] have somewhat spelled consistency across the regulations in Europe. The requirement of clear authorization to process personal data is a key requirement of GDPR. This would, in essence, refer to informing the subjects on the kind of data collected, and use made from it, and who can access them in the mHealth applications. They also allow individuals to access their data, rectify errors, and request erasure under the "right to be forgotten". These rules allow individuals more authority, while, technically and operationally speaking, app creators must be in a position to ensure that their products do not breach such rights [37]. Overall GDPR is an especially important regulation for national mobile EHRs to ensure privacy and data protection. It applies to any organization processing personal data of EU residents, including governments.

A more recent initiative is the EHDS regulation. While the GDPR provides general legislation across all sectors, the EHDS specifically focuses on healthcare data. This regulation will establish legal rules for managing health data (primary use) across Member States and promoting healthcare research (secondary use) [38]. The objective is to create a unified framework for data interoperability among European countries, ensuring the security of patient information.

Similarly in USA, Health Insurance Portability and Accountability Act (HIPAA) [39] is a U.S. regulation enacted in 1996 that sets national standards for protecting sensitive patient health information. It mandates the secure handling of personal health data by healthcare providers, insurers, and other entities that process these data. Key components of HIPAA include the Privacy Rule, which govern the use and disclosure of health information, and the Security Rule, which establishes safeguards to protect electronic health records. HIPAA also gives patient's rights over their data, including the ability to access their health records and request corrections.

5.2 Infrastructure development

When developing a national mobile EHR system, a lot of factors need to be considered. First of all, there is a need for centralized or distributed servers that need to be scalable as they are needed on a national level. Additionally, the infrastructure should have high-speed internet connectivity since most users on their smartphone will request documents, online sessions, etc., and the delivery of the services should be extremely fast. Moreover, the availability of the servers must be at 100%, but this is not realistic since there are malicious attacks and maintenance time. Nevertheless, these durations should be minimized by following best practices for defending against cyber-attacks and mechanisms to still have available services on maintenance [40].

An alternative to developing the infrastructure of such a system is cloud computing. As we discussed in Section 3.3, governments may consider using cloud services for storage, computing, and backup solutions. Due to legislation and security issues, it is not advised to have the entirety of the system on the cloud because it is such a critical infrastructure; however, it can be used as complementary. Additionally, a benefit of cloud computing is that the management of the servers is not the responsibility of the customer, which means countries that do not have the infrastructure or the experts to maintain it, can take advantage of cloud computing.

5.3 Stakeholder engagement

In a national mobile EHR system, there are numerous and diverse stakeholders, including healthcare professionals, citizens, hospitals, governmental bodies, smartphone companies, and others. It is crucial for private hospitals to collaborate and not be operating as isolated data silos. This cooperation ensures that healthcare professionals can access a citizen's medical history, even if the citizen visits a different hospital. Smartphone companies also play a critical role, as they are responsible for distributing mobile applications. Any software bugs on their devices could potentially compromise the security of the application.

The most important stakeholders are the citizens and healthcare professionals, who are the primary end users of the mobile application [41]. Before launching the app, it is essential to conduct workshops and webinars to introduce the app, train healthcare providers, and gather their feedback. This process helps identify any errors, streamline workflows, and highlight needed features. Additionally, we must consider that some citizens may have concerns about security or find the application challenging to use. Public awareness campaigns can help educate citizens about the benefits of a mobile app connected to the national EHR system, encouraging them to engage with and utilize the technology.

Figure 5 summarizes they key components that are essential to implementing a national mobile EHR system. If any of these components is missing, then the implementation will fail either because it does not comply with the legislation, or the resources are insufficient for the population, or the application was designed without being focused to tackle the problems the end users want.

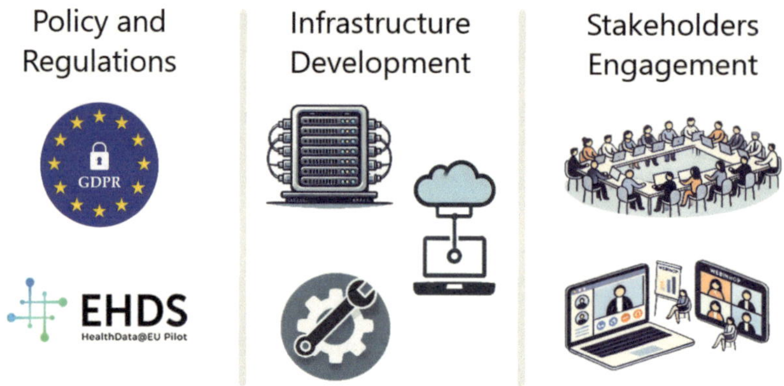

Figure 5.
Implementation strategies for a national mobile EHR system. Ensuring compliance with regulations such as GDPR is crucial for maintaining data privacy and protection. A reliable system also requires a robust infrastructure that integrates both physical and cloud-based solutions. Continuous communication with stakeholders is essential to align the system with user needs and expectations, ensuring its success and widespread adoption, and continuous communication with stakeholders is also vital to ensure the system meets user needs and expectations.

6. Selected EU country-specific mobile EHR implementations

In this section, we present examples of national EHR systems developed by various countries, organized alphabetically for clarity. Each country's solution is described in terms of its application, key initiatives, challenges faced, and achievements. Identifying a mobile EHR solution for each country can be a challenging task. To streamline this process, we focused our screening on a selection of prominent European countries. Our research methodology followed a structured, multi-step approach to ensure thorough coverage. First, we searched Google Scholar for publications detailing pilots or reviews of national mobile EHR applications in these countries. If this did not provide sufficient results, we turned to the Google search engine, using the query "Country mobile EHR," and reviewed official government websites. If these efforts were unsuccessful, we proceeded to check the Google Play Store, for applications published by the national eHealth authority.

6.1 France

"Mon Espace Santé" is a versatile mobile app designed to empower citizens by consolidating their health data in a single, accessible location [42]. This platform facilitates the sharing of medical records with trusted entities such as healthcare professionals, facilities, and approved digital services. Its four primary functionalities include managing and sharing medical records, scheduling patient appointments, a secure health messaging service, and an application catalog that features government-selected apps compatible with "Mon Espace Santé."

Drawing on their experience, the developers of "Mon Espace Santé" recognize a growing trend within healthcare systems to prioritize patient-centric approaches and meet patient expectations more effectively. The tools provided by the app have been instrumental in enhancing communication between patients and healthcare providers, strengthening their relationships. Moreover, engagement metrics show promising user interaction, with 30% of users returning monthly. This has led to more efficient management of medical documents, evidenced by the fact that the volume of documents stored has doubled in just a few months, surpassing years of accumulation in traditional systems [42].

6.2 Germany

Germany's Elektronische Patientenakte (ePA), launched in 2021, marks a significant advancement in the nation's digital health strategy [43]. This system empowers patients to actively manage their health data by allowing them to decide which healthcare providers can access specific information. Such autonomy has led to increased patient engagement, with users reporting a better understanding of their health conditions.

Complementing the ePA, the German government's "Act to Improve Healthcare Provision through Digitalisation and Innovation" introduces initiatives like "apps on prescription" and solutions to facilitate the use of video consultations [44]. This progressive approach has significantly benefited German citizens, enhancing their quality of life. By "apps on prescription," we mean that for specific diseases, the government has developed or approved specialized apps to assist in disease management. This groundbreaking initiative enables some medications to be supplemented or replaced by non-pharmacological interventions, all integrated within the ePA application.

6.3 Portugal

SNS24 is an integral service provided by the Portuguese National Health Service, "Serviço Nacional de Saúde" (SNS). It offers a combination of telephone and online support, aiding citizens, and their families in navigating both clinical and non-clinical aspects of the healthcare system. This service is invaluable for those seeking advice on acute, non-emergency health issues, enabling them to manage health-related concerns without the need to visit primary care units or hospitals. The array of services includes evaluation and counseling for non-emergency acute illnesses, fast and reliable information on assorted topics, and assistance with scheduling appointments in primary care, as well as support following an HIV self-test [45].

User acceptance testing, particularly with older adults, has highlighted areas for improvement and opportunities for autonomy. While many elderly users are capable of installing mobile applications, they often encounter difficulties with processes like registering on the SNS24 app [46]. However, with minimal assistance, they can successfully complete registration. This insight underscores the importance of tailoring user experiences to meet the needs of specific groups, such as the elderly, who may benefit from additional cues and interactive feedback. A notable enhancement could be the integration of voice assistance within the app, facilitating easier communication for users who find vocal interaction more accessible than typing or reading on small screens.

6.4 Spain

Spain's La Meva Salut is a personal health portal provided by the Catalan Health Service, offering a comprehensive mobile app that allows citizens to securely access their health information directly from their smartphones. Through the app, users can view medical records, lab results, vaccination histories, and medication plans, as well as manage appointments and communicate with healthcare professionals *via* secure messaging [47]. The integration of La Meva Salut with the Catalan Electronic Health Record system ensures that healthcare professionals have up-to-date information during clinical encounters, enhancing continuity and quality of care across different settings.

6.5 Sweden

Sweden's 1177 Vårdguiden is a national healthcare portal that plays a pivotal role in the country's digital health ecosystem. The platform offers a comprehensive mobile app that provides citizens with convenient access to personal health services directly from their smartphones [48]. Through the app, users can view their medical records, lab results, and vaccination histories. They can also book appointments, request prescription renewals, and receive personalized health advice.

The mobile app has significantly enhanced healthcare accessibility, especially for those living in remote or rural areas where healthcare facilities may be less accessible. By enabling secure communication between patients and healthcare providers, the app facilitates timely medical consultations and follow-ups [48]. Moreover, the 1177 Vårdguiden app integrates with Sweden's national electronic health records system, ensuring that healthcare providers have up-to-date information during clinical encounters [49].

6.6 United Kingdom

The NHS App is a simple and secure way to access a range of NHS services on your smartphone or tablet. The services include checking symptoms using NHS111 online, viewing your general practice medical record, booking and managing appointments, ordering repeat prescriptions, registering as an organ donor, and sharing national data securely.

The ecological study on the NHS App reveals several key insights into its adoption and impact. Primarily, the study highlights an inequality in app adoption across different demographic groups, influenced by factors related to both patients and providers. The use of the Interrupted Time Series (ITS) design allowed for an analysis of trends over time, particularly in relation to COVID-19, though it did not fully account for other rapid changes in healthcare delivery that could affect results. Despite a strong initial uptake, sustained use of the app remains a challenge, underscoring the need for further research to understand the barriers to continued engagement and the overall effectiveness of the app in enhancing health outcomes [50].

7. Best practices, challenges, and future trends

The integration of mobile Electronic Health Records (EHRs) into national healthcare systems is revolutionizing how health services are accessed and managed. This section delves into the benefits, challenges, and future prospects of national mobile EHR implementations. We will outline best practices to enhance user experience, discuss obstacles to successful deployment, and highlight emerging trends shaping the future of healthcare.

7.1 Best practices

As we synthesize the information gathered from the previous sections, we highlight best practices for national mobile EHR implementation in **Figure 6**. All these best practices can be applied in the services we listed in **Figure 1** to improve the flow, usability, and overall user experience of the user.

The first best practice to emphasize is the development of patient-centric solutions, placing citizens at the core of the system. Integrating features such as the digital wallet, teleconsultation, and accessibility options will streamline health management and enhance the overall quality of life for citizens [51, 52]. Additionally, the collection of high-quality data will improve the effectiveness of medical interventions and analytics, offering deeper insights to patients [53].

Another key best practice is strengthening the communication between doctors and patients through services such as secure messaging and online appointments. Furthermore, implementing multi-layered security measures is essential to ensure that sensitive data are only accessible to authorized individuals. Ensuring the app is equipped with accessibility features is crucial so that it can be used by everyone without restrictions. Lastly, the system must be interoperable with other platforms to remain future proof and capable of integrating additional functions that may rely on external systems.

7.2 Challenges

Moving on, we have identified numerous challenges that must be addressed to successfully implement a national mobile EHR application, as presented in **Table 4**. In this section, we briefly describe the main challenges of implementing an EHR system.

One of the most significant challenges is ensuring privacy and data security. The sensitive nature of health data makes it a prime target for cyberattacks, necessitating robust security measures. Implementing multi-layered security protocols, including encryption, secure authentication methods such as multi-factor authentication, and regular security audits, is essential to protect against data breaches. Countries are generally adopting comprehensive data protection regulations and guidelines that mandate strict compliance from all stakeholders involved [54]. Balancing the need for accessible health information with stringent privacy protections requires continuous efforts to update security practices in line with emerging threats and technological advancements.

Furthermore, achieving interoperability among diverse healthcare systems remains a significant technical hurdle. The lack of standardized data formats and communication protocols can lead to fragmented information silos, impeding the seamless exchange of patient data across different platforms and institutions. To address this, countries are increasingly adopting international interoperability

Figure 6.
Best practices for national mobile EHR solutions.

Challenges of mobile EHR systems			
Technical	**Political**	**Economic/Social**	**Legislation**
Multiple standards	Define common standards	Cultural and religious barriers	Comply with laws
Multi-layer security system	Lack of legal framework	Technological understanding	Ownership of data
Interoperability with other systems	Lack of shared technical standards	Accessibility to minorities	Mechanics for deletion of data
Collaboration with technology companies	Management of secondary use of data	Lack of digital skills	Authorized access to shared medical data
Availability	Liability of EHRs	High costs	

Table 4.
Challenges of mobile EHR systems for technical, political, economic, and legislation sectors.

standards like HL7 FHIR and promoting the use of open APIs. Collaborative initiatives among governments, healthcare providers, and technology developers aim to establish common frameworks and guidelines that facilitate efficient data sharing [12, 29, 36, 38, 39]. By focusing on interoperability, healthcare systems can enhance the continuity of care, reduce medical errors, and improve overall patient outcomes.

Politically, a major challenge is creating a cohesive legal framework that balances national sovereignty with cross-border data sharing. While the benefits of a unified mobile EHR system include improved patient care and the ability to access health data internationally, political tensions may arise regarding data ownership, governance, and the liability of healthcare providers. Furthermore, ensuring collaboration between public and private healthcare providers is essential for the system's success. This requires a coordinated effort between government bodies, technology companies, healthcare institutions, and other stakeholders to build a system that is both scalable and sustainable.

On the social and economic front, disparities in digital literacy and access to healthcare technology present significant barriers. For a national mobile EHR system to succeed, citizens need to be educated on how to use these platforms effectively. Public awareness campaigns, training programs, and ongoing support for healthcare professionals and patients alike are essential to bridge the digital divide. Additionally, economic challenges, such as the high costs associated with developing and maintaining such systems, must be addressed. Governments and healthcare providers need to find ways to allocate sufficient resources to ensure that these systems are accessible, secure, and effective for all users.

7.3 Future trends

As we look ahead, several emerging trends are poised to redefine how healthcare services are delivered and experienced. These trends, as shown in **Figure 1**, focus on enhancing patient engagement, personalizing care, advancing data management and security, and leveraging technological innovations to improve efficiency and outcomes. By embracing these developments, healthcare systems can address current challenges and set new standards for patient care in the digital age (**Table 5**).

In recent years, Europe has strategically funded numerous projects with clearly defined milestones to prepare for this future. For example, POTENTIAL [55] is a

Future trends of mobile EHR systems		
Patient engagement and personalization	**Data management and security**	**Technology-driven innovations**
Digital wallet	Seamless data sharing	5G+ technologies
Translated patient summaries	Blockchain technology	AI-powered decision support
Personalized care	Interoperable health records	Telehealth innovations and remote patient monitoring

Table 5.
Emerging mHealth features that are expected to be integrated into healthcare system in the coming years.

major initiative, engaging multiple member states to develop a reference implementation for a national Digital Wallet. Another project, PATHeD [56], addresses the issue of translated patient summaries. Its pilots demonstrate that citizens are eager to utilize this feature, especially when traveling abroad. Additionally, the xShare [57] initiative aims to revolutionize the seamless sharing of medical data, allowing individuals to share their health information with authorized parties at the touch of a button. Finally, the Xt-EHR project [58] is dedicated to establishing guidelines for the development of a comprehensive, interoperable, and secure EHR system that promotes smooth connections among healthcare providers within Member States, ensuring compliance with the EHDS regulation. This initiative will drive the integration of mHealth solutions to enhance accessibility and patient engagement.

Personalized care is another emerging trend. While healthcare professionals may have limited capacity to offer individualized care, mHealth applications, specifically designed for targeted groups, are making this possible. For instance, virtual reality mobile applications can create calming environments for dementia patients [59] or offer intrahospital experiences to bedridden patients [60], while other interventions provide mental, cognitive, and physical support for cancer patients [61]. These tools also enable remote monitoring, allowing healthcare providers to adjust telehealth interventions based on patient progress.

Blockchain technology is also showing promise in securely transferring medical data. By providing decentralized and tamper-proof data storage, blockchain enhances the security and privacy of EHRs. Applications already exist that allow patients to control access to their health summaries, and the feasibility of this technology has been successfully tested [62]. Moreover, as mentioned earlier, 5G technology is transforming the potential of EHRs with ultra-fast, low-latency connectivity. This allows for real-time data exchange and access to comprehensive patient records, even in remote areas.

Lastly, AI-driven solutions represent one of the most transformative trends shaping the future of healthcare. Their potential extends beyond the obvious, encompassing everything from intelligent chatbots and advanced visual aids to sophisticated diagnostic support systems. With the ability to seamlessly integrate into mobile health coaching platforms, AI excels at gathering and analyzing datasets, uncovering actionable insights, and delivering personalized guidance through intuitive, human-like interactions. This capacity for real-time, data-driven decision-making not only enhances the quality of care but also empowers both patients and healthcare professionals to make informed decisions with greater efficiency.

8. Conclusions

In conclusion, nations are increasingly adopting mobile EHR solutions, transitioning from outdated paper-based methods to modern, digital health record management. To successfully implement these systems, it is crucial to focus on practical steps such as investing in robust technological infrastructure, ensuring interoperability through standardized protocols, and engaging stakeholders, including patients, healthcare providers, and policymakers, in the development process. This shift holds significant potential to enhance quality of life by providing instant access to health records, integrating healthcare services into cloud-based systems, and improving communication between patients and healthcare providers. For patients, mobile EHRs offer tangible benefits such as greater control over their health information, improved access to healthcare services, and personalized care through features such as teleconsultations and health monitoring. Healthcare providers benefit from streamlined workflows, reduced administrative burdens, and enhanced ability to make informed clinical decisions due to real-time access to comprehensive patient data.

As we move forward, it is anticipated that citizens will enhance their digital skills, enabling them to fully leverage mobile EHR applications. With effective education and outreach, these tools can become accessible and beneficial to all. Implementing training programs for both patients and healthcare professionals is essential to maximize the adoption and effective use of mobile EHR systems. Additionally, addressing privacy and security concerns through strict compliance with data protection regulations builds trust among users. However, the success of mobile EHR systems relies on ongoing public education, strategic promotion, and the maintenance of public trust through adherence to the highest standards of transparency, data security, and ethical development.

Our literature review indicates that the development and management of mobile EHR systems is a complex endeavor, presenting numerous challenges that require collaborative efforts to overcome. Nations can greatly benefit from sharing insights and lessons learned in this area. By establishing international collaborations and sharing best practices, countries can accelerate the effective implementation of mobile EHR systems. We are now in an era where rapid technological innovation is reshaping the healthcare landscape, and governments must adapt by delivering services through these new digital platforms to ensure accessibility for most citizens. This can be achieved by focusing on practical implementation strategies and highlighting the benefits of patient outcomes, efficiency, and cost savings, governments, and healthcare organizations can drive the adoption of mobile EHR systems that truly benefit both patients and providers. Crucially, this transformation should be supported by mechanisms that enhance healthcare delivery, such as instant access to medical data, online consultations, AI-driven coaching systems, smart interventions, and the integration of IoT-enabled medical devices.

Acknowledgements

This work was funded partly by the projects: *Deployment of generic cross border eHealth services in Cyprus*, funded by EU Recovery and Resilience plan for Cyprus, Apr. 2022–Sep. 2025; *PATHED: Enabling Patient Access to their Health Data*, EU4H-2021-DGA-MS-IBA-02 Type of Action: EU4H-PJG, Jan. 2023–Jun. 2024; *POTENTIAL:*

Pilots for European Digital Identity Wallet, Digital Europe Programme, Apr. 2023–Mar. 2025; *Xt-EHR: Extended EHR@EU Data Space for Primary Use*, EU4H-2022-JA-IBA (Joint Actions 2022), Oct. 2023–Mar. 2026; and *xShare: Expanding the European EHRxF to share and effectively use health data within the EHDS*, HORIZON-HLTH-2023-IND-06 European Health and Digital Executive Agency, Dec. 2023–Nov. 2026.

Author details

Theodoros Solomou[1]*, Christos N. Schizas[2] and Constantinos S. Pattichis[1]

1 University of Cyprus, Nicosia, Cyprus

2 National eHealth Authority, Nicosia, Cyprus

*Address all correspondence to: solomou.theodoros@gmail.com

References

[1] Istepanian R, Laxminarayan S, Pattichis CS. M-health: Emerging mobile health systems. New York: Springer Science & Business Media; 2007

[2] European Parliament. EU Health Data Space: more efficient treatments and life-saving research. Brussels: European Parliament News; 2024. Available from: https://www.europarl.europa.eu/news/en/press-room/20240419IPR20573/eu-health-data-space-more-efficient-treatments-and-life-saving-research [Accessed: August 27, 2024]

[3] European Commission. European Health Data Space Regulation. Brussels: EUR-Lex; 2024 . Available from: https://eur-lex.europa.eu/legal-content/EN/TXT/HTML/?uri=CELEX:52022PC0197 [Accessed: August 27, 2024]

[4] OpenAI. ChatGPT Can Now See, Hear, and Speak. San Francisco: OpenAI. Available from: https://openai.com/index/chatgpt-can-now-see-hear-and-speak/; 2023 [Accessed: August 6, 2024]

[5] Juang WC, Hsu MH, Cai ZX, Chen CM. Developing an AI-assisted clinical decision support system to enhance in-patient holistic health care. PLoS One. 2022;**17**(10):e0276501

[6] International Patient Summary. International Patient Summary. 2024. Available from: https://international-patient-summary.net/ [Accessed: August 6, 2024]

[7] Bruthans J, Jiráková K. The current state and usage of European electronic cross-border health services (eHDSI). Journal of Medical Systems. 2023;**47**(1):21

[8] Carrillo de Albornoz S, Sia KL, Harris A. The effectiveness of teleconsultations in primary care: systematic review. Family Practice. 2022;**39**(1):168-182

[9] Juneau CE, Briand AS, Collazzo P, Siebert U, Pueyo T. Effective contact tracing for COVID-19: A systematic review. Global Epidemiology. 2023;**5**:100103

[10] Sharif A, Ranzi M, Carbone R, Sciarretta G, Marino FA, Ranise S. The eIDAS regulation: a survey of technological trends for European electronic identity schemes. Applied Sciences. 2022;**12**(24):12679

[11] Boissin C, Blom L, Wallis L, Laflamme L. Image-based teleconsultation using smartphones or tablets: qualitative assessment of medical experts. Emergency Medicine Journal. 2017;**34**(2):95-99

[12] European Union. Regulation (EU) 2021/123 on a framework for the issuance, verification and acceptance of interoperable COVID-19 vaccination, testing and recovery certificates (eIDAS 2.0). Official Journal of the European Union, Brussels: Official Journal of the European Union. 2021:1-30

[13] Lakhala M, Benslimane M, Tmimi M, Ibriz A. Architecture of real-time patient health monitoring based on 5G technologies. Advances in Science, Technology and Engineering Systems Journal. 2021;**6**(2):351-358

[14] Li X, Zhao L, Xu T, Shi G, Li J, Shuai W, et al. Cardiac telerehabilitation under 5G internet of things monitoring: a randomized pilot study. Scientific Reports. 2023;**13**(1):18886

[15] Majumder S, Deen MJ. Smartphone sensors for health monitoring and diagnosis. Sensors. 2019;**19**(9):2164

[16] Tamura T. Current progress of photoplethysmography and SPO2 for health monitoring. Biomedical Engineering Letters. 2019;**9**(1):21-36

[17] Aghdam ZN, Rahmani AM, Hosseinzadeh M. The role of the Internet of Things in healthcare: Future trends and challenges. Computer Methods and Programs in Biomedicine. 2021;**199**:105903

[18] Yasar K, Gillis AS. What is Internet of Things (IoT)?—Definition from IoT Agenda. TechTarget. 2024. Available from: https://www.techtarget.com/iotagenda/definition/Internet-of-Things-IoT [Accessed: August 6, 2024]

[19] Fisher AJ. Toward a dynamic model of psychological assessment: Implications for personalized care. Journal of Consulting and Clinical Psychology. 2015;**83**(4):825

[20] Terblanche N, Molyn J, De Haan E, Nilsson VO. Coaching at Scale: Investigating the Efficacy of Artificial Intelligence Coaching. International Journal of Evidence Based Coaching & Mentoring. 2022;**20**(2):20-36

[21] Srinivasu PN, Sandhya N, Jhaveri RH, Raut R. From blackbox to explainable AI in healthcare: existing tools and case studies. Mobile Information Systems. 2022;**2022**(1):8167821

[22] Amann J, Blasimme A, Vayena E, Frey D, Madai VI. Precise4Q Consortium. Explainability for artificial intelligence in healthcare: a multidisciplinary perspective. BMC Medical Informatics and Decision Making. 2020;**20**:1-9

[23] Aceto G, Persico V, Pescapé A. Industry 4.0 and health: Internet of things, big data, and cloud computing for healthcare 4.0. Journal of Industrial Information Integration. 2020;**18**:100129

[24] Awaisi KS, Hussain S, Ahmed M, Khan AA, Ahmed G. Leveraging IoT and fog computing in healthcare systems. IEEE Internet of Things Magazine. 2020;**3**(2):52-56

[25] Benson T, Grieve G. Principles of Health Interoperability: SNOMED CT, HL7 and FHIR. London: Springer; 2016

[26] Saripalle RK. Leveraging FHIR to integrate activity data with electronic health record. Health and Technology. 2020;**10**(1):341-352

[27] Müller ML, Ückert F, Bürkle T, Prokosch HU. Cross-institutional data exchange using the clinical document architecture (CDA). International Journal of Medical Informatics. 2005;**74**(2-4):245-256

[28] Data.Gov.ie. Master Value Catalogue. 2014. Available from: https://datacatalogue.gov.ie/dataset/master-value-catalogue

[29] International Organization for Standardization. Health informatics—Detailed clinical models, characteristics and processes [ISO 13972:2015]. Geneva: International Organization for Standardization; 2015

[30] Label2Enable: Promoting the adoption of trustworthy digital health solutions. Available from: https://label2enable.eu/

[31] Essén A, Scandurra I, Gerrits R, Humphrey G, Johansen MA, Kierkegaard P, et al. Patient access to electronic health records: differences across ten countries. Health Policy and Technology. 2018;**7**(1):44-56

[32] Sahi MA, Abbas H, Saleem K, Yang X, Derhab A, Orgun MA, et al.

Privacy preservation in e-healthcare environments: State of the art and future directions. IEEE Access. 2017;**6**:464-478

[33] Vovk O, Piho G, Ross P. Methods and tools for healthcare data anonymization: a literature review. International Journal of General Systems. 2023;**52**(3):326-342

[34] Braunstein ML. Health informatics on FHIR: How HL7's new API is transforming healthcare. Cham, Switzerland: Springer International Publishing; 2018

[35] Hardt D, editor. The OAuth 2.0 Authorization Framework. RFC 6749. Fremont: Internet Engineering Task Force (IETF); 2012

[36] Regulation P. Regulation (EU) 2016/679 of the European Parliament and of the Council. Regulation (EU). 2016;**679**:2016

[37] Politou E, Michota A, Alepis E, Pocs M, Patsakis C. Backups and the right to be forgotten in the GDPR: An uneasy relationship. Computer Law & Security Review. 2018;**34**(6):1247-1257

[38] Terzis P, Santamaria Echeverria OE. Interoperability and governance in the European Health Data Space regulation. Medical Law International. 2023;**23**(4):368-376

[39] Health Insurance Portability and Accountability Act of 1996, Pub. L. No. 104-191, 110 Stat. 1936 (1996)

[40] Toeroe M, Tam F, editors. Service Availability: Principles and Practice. Chichester: John Wiley & Sons; 2012

[41] Domecq JP, Prutsky G, Elraiyah T, Wang Z, Nabhan M, Shippee N, et al. Patient engagement in research: a systematic review. BMC Health Services Research. 2014;**14**:1-9

[42] Hayar S, Mériade L, Duvauchel AG. Diffusion of e-health tools and services to their users: the example of "Mon Espace Santé" in France. Gestion et Management Public. 2024;**02**:Ih-XXIh

[43] Bundesministerium für Gesundheit. Elektronische Patientenakte. Bundesgesundheitsministerium.de. 2021. Available from: https://www.bundesgesundheitsministerium.de/elektronische-patientenakte [Accessed: October 10, 2023]

[44] Sauermann S, Herzberg J, Burkert S, Habetha S. DiGA—A chance for the German healthcare system. Journal of European CME. 2022;**11**(1):2014047

[45] Digital Health Europe. SNS24. Digital Health Europe; 2023. Available from: https://digitalhealtheurope.eu/covid-19/sns24/ [Accessed: July 28, 2024]

[46] Damas C, Pereira D, Gil H. Handled application, life made easy?: SNS 24 application: evaluation of its use with the elderly. In: 2023 18th Iberian Conference on Information Systems and Technologies (CISTI). Aveiro: IEEE; 2023. pp. 1-6

[47] Servei Català de la Salut. La Meva Salut. Barcelona: Generalitat de Catalunya. 2021. Available from: https://catsalut.gencat.cat/ca/serveis-sanitaris/la-meva-salut/index.html; [Accessed: October 10, 2023]

[48] Hägglund M, Blease C, Scandurra I. Mobile access and adoption of the Swedish national patient portal. In: Integrated Citizen Centered Digital Health and Social Care. Amsterdam: IOS Press; 2020. pp. 82-86

[49] Ulfvarson J. National report: Sweden. Ageing and Technologies. 2020;**5**(6):123-137

[50] Sukriti KC, Tewolde S, Laverty AA, Costelloe C, Papoutsi C, Reidy C, et al. Uptake and adoption of the NHS App in England: an observational study. British Journal of General Practice. 2023;**73**(737):e932-e940

[51] Karacic Zanetti J, Nunes R. To wallet or not to wallet: The debate over digital health information storage. Computers. 2023;**12**(6):114

[52] Yusof K, Neoh KH, bin Hashim MA, Ibrahim I. Role of teleconsultation in moving the healthcare system forward. Asia Pacific Journal of Public Health. 2002;**14**(1):29-34

[53] Sami O, Elsheikh Y, Almasalha F. The role of data pre-processing techniques in improving machine learning accuracy for predicting coronary heart disease. International Journal of Advanced Computer Science and Applications. 2021;**12**(6). DOI: 10.14569/IJACSA.2021.0120695

[54] Atadoga A, Farayola OA, Ayinla BS, Amoo OO, Abrahams TO, Osasona F. A comparative review of data encryption methods in the USA and Europe. Computer Science & IT Research Journal. 2024;**5**(2):447-460

[55] Digital Identity Wallet. Facilitating Digital Identities Across Europe. Digital Identity Wallet. 2024. Available from: https://www.digital-identity-wallet.eu/

[56] Solomou T, Canciu IC, Christodoulou M, Savva P, Yiasemi C, Antoniou Z, et al. Empowering Citizens Through Translated Patient Summary Access and Sharing in Digital Health Ecosystems. Studies in Health Technology and Informatics. 2024;**316**:497-501

[57] XShare Project. Empowering Citizens in Digital Health Ecosystems. XShare Project. 2024. Available from: https://xshare-project.eu/

[58] XT-EHR: Cross-Border Management of Individual Patient Summaries in Digital Health Ecosystems. 2024. Available from: https://www.xt-ehr.eu/

[59] Matsangidou M, Solomou T, Frangoudes F, Ioannou K, Theofanous P, Papayianni E, et al. Affective out-world experience via virtual reality for older adults living with mild cognitive impairments or mild dementia. International Journal of Environmental Research and Public Health. 2023;**20**(4):2919

[60] Matsangidou M, Frangoudes F, Solomou T, Papayianni E, Pattichis C. Free of walls: Participatory design of an out-world experience via virtual reality for dementia in-patients. In: Adjunct Proceedings of the 30th ACM Conference on User Modeling, Adaptation and Personalization. New York: Association for Computing Machinery; 2022. pp. 326-334

[61] Matsangidou M, Kassianos AP, Papaioannou D, Solomou T, Krini M, Karekla M, et al. Virtual painkillers: Designing accessible virtual reality experiences for helping cancer patients manage pain at home. In: CHI Conference on Human Factors in Computing Systems Extended Abstracts. New York: Association for Computing Machinery; 2022. pp. 1-9

[62] Hölbl M, Kompara M, Kamišalić A, Nemec ZL. A systematic review of the use of blockchain in healthcare. Symmetry. 2018;**10**(10):470

Chapter 5

Evaluation of U-Net Algorithm Precision for Segmentation and Classification of Breast and Lung Images on CBIS-DDSM and LIDC-IDRI Datasets

Ajna Fetić and Nevzudin Buzađija

Abstract

Cancer is the leading disease in the world in terms of increasing numbers of new patients and deaths annually, making it one of the most formidable diseases today. Lung cancer and breast cancer are widely acknowledged as two of the most prevalent forms of cancer, both falling under the carcinoma subtype. Early detection of these cancers is particularly crucial for patient outcomes. With cancer being a longstanding health challenge, comprehensive datasets containing essential information for cancer prediction and diagnosis are now available. Radiologists primarily rely on Computed Tomography (CT) scans for cancer diagnosis. However, the escalating demand for CT scans and their interpretation has led to radiologists prioritizing throughput over the quality of readings. To address this challenge, Computer-Aided Diagnosis (CAD) systems have been introduced into medical practice. This research paper delves into evaluating the accuracy of the implemented U-Net algorithm for segmentation and classification on two prominent medical datasets: Curated Breast Imaging Subset of DDSM (CBIS-DDSM) and Lung Image Database Consortium and Image Database Resource Initiative (LIDC-IDRI). The primary objective of the study is to assess the algorithm's performance in accurately identifying and classifying Regions of Interest (ROIs) in breast and lung medical images.

Keywords: U-Net, segmentation, classification, breast images, lung images, CBIS-DDSM, LIDC-IDRI, medical dataset, deep learning, algorithm evaluation

1. Introduction

Early detection is key when it comes to survival in cases of lung, breast, or any other type of cancer. Cancer has been around for a long time, and unfortunately, the number of cases has only grown. Fifty years ago, diagnosing and predicting cancer relied heavily on one medical expert running tests like mammograms, ultrasounds, and MRIs. As explained in Ref. [1], an ultrasound works by sending sound waves through the skin to check what is happening inside the body. A device called a transducer is placed on the

IntechOpen

skin, and it sends out sound waves that bounce off tissues, creating echoes. These echoes are turned into grayscale images using binary values. The same paper [1] also discusses PET scans, which are often used to spot early signs of cancer, heart disease, or brain disorders by injecting radioactive tracers that help highlight diseased cells. For example, injecting F-fluorodeoxyglucose allows doctors to see exactly where a tumor is located. Another tool, dynamic MRI, is used to examine blood vessels and is often linked to detecting metastasis, while mammography uses X-rays to get a better look at tissue.

As the years passed, diagnosing cancer became more complex. Doctors now have to sift through a wide range of tests and variables to figure out if cancer is benign or malignant, all while trying to avoid personal bias. Since the human mind is not perfect and can lack objectivity, the need for automation became clear. For instance, in mammography, doctors often have to analyze a large volume of images, which can lead to errors.

In recent years, a lot of research has focused on how machine learning can assist in cancer diagnosis and prediction. The reason this has gained attention is that machine learning, a branch of artificial intelligence, uses algorithms to sift through massive datasets, "learn" from the data and produce results. We'll explore different types of machine learning later, but overall, it is clear that machine learning is a major step forward in diagnostic medicine. In cancer prognosis, where there is often so much data to analyze, it can help reduce human error and speed up diagnoses. The aim is not to replace doctors but to support them, making the diagnostic process quicker and more accurate.

In this study, our primary objective is to explore the performance and accuracy of the U-Net architecture on a model designed specifically for car image segmentation when it is used on a lung and breast cancer dataset. The algorithm used in this paper was created on the U-Net architecture for the Carvana dataset and is used for car/vehicle segmentation on images. Here we want to see if it is possible to reuse a solution for a different problem—lung cancer segmentation—and see how this reused solution performs in comparison to algorithms specifically developed for lung cancer segmentation and classification, namely Pro-CAN, DeepLung, Local-Global and NASlung.

Computer-Aided Diagnosis (CAD) systems have emerged as powerful tools in the field of medical imaging, aiding healthcare professionals in the detection and diagnosis of various diseases, including cancer. These systems leverage advanced algorithms and machine learning techniques to analyze medical images, extract relevant features, and provide insights that aid in accurate and efficient diagnosis.

2. Related work

To provide a more comprehensive evaluation, it is crucial to compare the U-Net architecture with other recent state-of-the-art deep learning models that have shown potential in medical imaging. Architectures like ResNet, DenseNet, and Transformer-based models have achieved notable results in tasks such as image segmentation and classification. These models offer enhanced feature extraction capabilities and deeper representations of image data, which can improve accuracy and precision in medical applications. While U-Net remains a powerful tool, especially for segmentation tasks, exploring these alternative architectures would help highlight its strengths and limitations. This comparison can further illustrate why U-Net was chosen for this study, especially given its simplicity and effectiveness in cases requiring precise localization of features, such as breast and lung imagery.

In the previous studies on lung nodule classification using the algorithms described in this paper, the authors address the challenging task of classifying diverse lung nodules as benign or malignant based on their shapes and sizes. The proposed methodology incorporates Residual Blocks for extracting local features, known for their efficacy in CNNs designed for nodule classification. Additionally, the authors employ Non-Local neural networks, incorporating Self-Attention layers, to capture global features by analyzing the density and structure of the nodules. The Non-Local Blocks utilize matrix multiplication to create non-linear interactions between spatial features, and the Softmax function generates attention masks for selective feature retention. The studies compare the proposed Local-Global Network with various baseline methods, including Densenet121, Resnet50, Resnet18, BasicResnet, AllAtn, and AllAtnBig [2]. These baselines involve pretrained architectures and variations, allowing for a comprehensive analysis. The results demonstrate the effectiveness of the Local-Global Network, leveraging both local and global features for improved lung nodule classification. The inclusion of Non-Local Blocks proves to be crucial, as evidenced by comparisons with baseline methods, showcasing the significance of incorporating global feature extraction in enhancing the accuracy of malignancy predictions [3].

In addition to the methodology presented in this study, several other approaches have been proposed to tackle the complexities of lung nodule classification. In paper [4], a 3D Convolutional Neural Network (3D-CNN) that incorporates a hybrid loss function combining focal loss and Dice coefficient is introduced to address the imbalance between benign and malignant nodules in training datasets. This approach focuses on improving classification performance in datasets with high variability in nodule sizes and shapes, highlighting the importance of tailored loss functions in enhancing model robustness.

In addition to technical performance, it is critical to address the ethical considerations surrounding the use of AI in healthcare. One major concern is patient data privacy. Ensuring that AI models are developed and deployed while maintaining the confidentiality of sensitive health data is paramount. Additionally, biases in training datasets are another ethical challenge. If certain patient demographics are underrepresented, the model may not perform equally well across diverse populations, potentially leading to biased diagnoses. The integration of ethical frameworks that safeguard patient privacy and ensure the fairness of AI models is necessary for their widespread acceptance and effectiveness in clinical practice.

In paper [5], a Multi-scale Dense Convolutional Neural Network (MD-CNN) is proposed, integrating multi-scale feature extraction with dense connections to improve the classification of pulmonary nodules. This method captures both fine-grained and coarse-grained features, leading to better handling of nodules with complex textures. The study demonstrates that integrating multi-scale features significantly enhances classification accuracy, especially for small and irregular nodules. In paper [6], a Two-Stream Network architecture is developed that combines a standard CNN with a Transformer-based model to capture both local and contextual information in lung nodules. This approach leverages the strengths of CNNs in local feature extraction and Transformers in capturing long-range dependencies, leading to improved classification outcomes. This study underscores the potential of hybrid architectures in achieving state-of-the-art performance in medical image classification tasks. In paper [7], a deep learning approach specifically designed to address the variation in nodule shapes and textures is introduced. The model, based on a multi-crop pooling strategy, enhances the sensitivity of the network to nodule variations by allowing multiple views of the nodules during training. This technique improved the network's ability to differentiate between benign and malignant nodules by focusing on the most informative regions within the images.

In paper [8], the effectiveness of multi-view CNNs for lung nodule classification is explored. By analyzing nodules from nine different views, this approach significantly improved the detection and classification of nodules with irregular shapes or those partially obscured by other lung structures. The multi-view strategy provided a more comprehensive analysis, allowing for better classification accuracy, particularly in complex cases where single-view analysis might fail. In paper [9], an end-to-end nodule detection and classification pipeline that utilizes a multi-task learning framework is proposed. By jointly learning the tasks of nodule detection and malignancy classification, the model achieved higher overall accuracy and robustness. The multi-task approach ensured that the features learned during nodule detection were directly applicable to classification, leading to a more cohesive and efficient model.

3. Research hypothesis

The central hypothesis in this research revolves around the belief that the algorithm initially created for the precise segmentation of vehicles within the Carvana dataset on the U-Net architecture, holds the potential to be a transformative force in the world of lung cancer detection. Our focal point is the intricate domain of CT scan images, where the shadows of lung cancer often linger, eluding early discovery. By harnessing the power of innovative image processing techniques and leveraging statistical metrics such as the F1 score, AUC, IoU, and Dice score, we anticipate identifying algorithms that exhibit superior performance in identifying lung nodules and segmenting them accurately. Furthermore, by developing a user-friendly application that enables individuals to access and interpret their CT scan results conveniently, we aim to bridge the gap between sophisticated algorithms and practical healthcare applications, improving patient outcomes and reducing the burden of lung cancer.

3.1 Possible results

H0: Accuracy of this U-Net implementation should be greater than 80%. Extremely High Precision in Nodule Detection: If our research aligns with this scenario, it signifies an alignment with our initial hypothesis. In this case, the U-Net algorithm may demonstrate exceptional precision in detecting nodules, heralding a change in basic assumptions in medical imaging.

H1: Predictive capabilities with cautions—Algorithm usage with advisory or limited nodule detection: In this outcome, our research suggests that the algorithm possesses predictive capabilities, albeit with some nuances. While it may not achieve the unparalleled precision anticipated in H0, it could still serve as a valuable tool for preliminary nodule identification. This outcome underscores the importance of nuanced interpretation and collaboration between artificial intelligence tools and healthcare professionals.

4. Datasets

4.1 LIDC/IDRI

The Lung Image Database Consortium and Image Database Resource Initiative (LIDC/IDRI) dataset stands as the most popular dataset in lung cancer detection and research. It combines a diverse collection of Computed Tomography (CT) scans, and this dataset has been important in advancing the development of machine learning algorithms and Computer-Aided Diagnosis systems for the early detection of lung cancer. With its unique combination of CT images and expert annotations, the LIDC/IDRI dataset offers a resource for scientists and medical professionals striving to improve the accuracy and efficacy of lung cancer diagnosis. LIDC-IDRI contains 1018 low-dose lung CTs from 1010 lung patients. The LIDC/IDRI dataset contains a variety of data relevant to lung cancer detection. It primarily consists of high-resolution CT scans, capturing the intricate details of the lung structures. These scans are obtained from a variety of patients, contributing to the diversity of cases encountered, from benign to malignant lung nodules. In total, the dataset contains thousands of CT scans, making it a robust resource for algorithm training and evaluation. An important feature of the LIDC/IDRI dataset is the annotation process it undergoes. Expert radiologists review each CT scan, marking and annotating lung nodules and other abnormalities. These annotations often involve a consensus among multiple radiologists, ensuring the reliability of the ground truth data. This consensus-based approach contributes to the dataset's credibility, making it an invaluable tool for algorithm development. Despite its strengths, the LIDC/IDRI dataset presents different challenges and variabilities. Lung nodules come in diverse shapes, sizes, and densities, reflecting the complexity of real-world clinical cases. This diversity is a double-edged sword, providing many opportunities for algorithm development but also demanding sophisticated solutions to manage this variability. Researchers must address these challenges to create algorithms that can generalize effectively across different nodule characteristics. One of the most significant advantages of the LIDC/IDRI dataset is its public availability. This accessibility has encouraged a wide range of research studies, resulting in numerous publications and innovations in the field of lung cancer detection. Researchers worldwide have leveraged this dataset to develop and benchmark machine learning models, contributing to advancements in early diagnosis and the overall management of lung cancer. **Figure 1** shows an example of nodule segmentation on the LIDC/IDRI dataset.

4.2 CBIS/DDSM

The Curated Breast Imaging Subset of the Digital Database for Screening Mammography (CBIS-DDSM) dataset is one of the most widely utilized datasets in breast cancer detection and research. This dataset comprises a comprehensive collection of mammographic images, playing a crucial role in advancing the development of machine learning algorithms and Computer-Aided Diagnosis systems aimed at early breast cancer detection. With its extensive collection of annotated mammograms, the CBIS-DDSM dataset serves as a valuable resource for researchers and medical professionals working to enhance the accuracy and effectiveness of breast cancer diagnosis.

CBIS-DDSM contains a large number of mammogram images that are carefully curated to support the training and evaluation of machine learning models. This dataset is a subset of the original DDSM dataset, which was developed to provide researchers with a high-quality resource for mammographic image analysis. The images in CBIS-DDSM are captured using various mammography systems, ensuring diversity in image quality and presentation. This diversity is crucial for developing robust algorithms capable of generalizing across different imaging conditions and patient demographics.

One of the key features of the CBIS-DDSM dataset is the detailed annotation provided by expert radiologists. Each mammogram in the dataset is meticulously labeled,

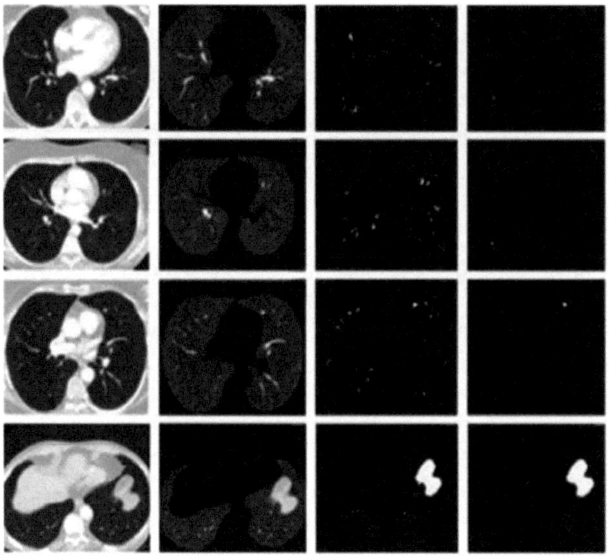

Figure 1.
Segmentation of the LIDC/IDRI dataset [10].

with annotations that include the location, size, and type of lesions, such as masses and calcifications. These annotations often come with corresponding BI-RADS (Breast Imaging-Reporting and Data System) scores, which provide standardized information on the likelihood of malignancy. The careful annotation process, often involving multiple expert reviews, ensures the reliability and accuracy of the data, making CBIS-DDSM an invaluable tool for developing and testing breast cancer detection algorithms.

The CBIS-DDSM dataset also presents various challenges and complexities, reflecting the real-world scenarios encountered in clinical practice. The mammograms in the dataset include a wide range of breast tissue densities, from fatty to extremely dense, which can affect the visibility of lesions. Additionally, the dataset includes both benign and malignant findings, adding to the complexity of developing algorithms that can accurately differentiate between the two. This variability in tissue density and lesion characteristics requires sophisticated image processing and machine learning techniques to ensure accurate detection and classification.

Despite these challenges, the CBIS-DDSM dataset is a powerful resource for the research community, particularly because of its public availability. The accessibility of this dataset has enabled a multitude of research studies, leading to significant advancements in the field of breast cancer detection. Researchers worldwide have utilized CBIS-DDSM to develop and benchmark machine learning models, contributing to improvements in early diagnosis and the overall management of breast cancer. The dataset's widespread use in the scientific community has resulted in numerous publications and innovations, further solidifying its role as a cornerstone in breast cancer research.

5. Data preprocessing

In this study, a comprehensive preprocessing pipeline was set up to prepare the datasets for lung and breast cancer detection. This process involved several key steps to ensure the data was consistent and ready for training machine learning models, especially for tasks like segmentation and classification. The preprocessing for lung cancer data started with isolating the lung regions from the CT scans in the LIDC-IDRI dataset. The original DICOM images were converted into NumPy arrays, making them easier to work with. Using the metadata associated with each scan, we created binary masks to segment the lung nodules from the lung regions, which was crucial for focusing on the areas where cancer might be present. Alongside the segmentation masks, we also extracted diagnostic information—whether each nodule was malignant (cancerous) or benign (non-cancerous). This information was used to create binary labels for classification, with "true" indicating cancerous nodules and "false" for non-cancerous ones. A new metadata file was then generated to store this information, and a fresh dataset was created that included the processed images, segmentation masks, and classification labels. To keep things consistent with the CBIS-DDSM dataset (which does not include cases without masses), we excluded any "clean" cases from our lung cancer dataset—meaning patients who had no lung nodules were left out. By focusing only on cases with nodules, we ensured that our dataset was suited for binary classification, which was necessary for the models to work properly.

For the CBIS-DDSM dataset, the preprocessing required similar steps. We started by creating a CSV file that mapped the file paths of the images and their corresponding masks. The dataset contained a mix of mammogram images, pre-generated binary masks, and DICOM images that only showed nodules. We carefully identified and categorized these files to make sure that the images and masks were paired correctly. Once everything was organized, we created a new dataset that matched the structure of our lung cancer dataset. During this process, we segmented the breast regions from the mammograms to remove any background noise. The mammograms were labeled with MLO and CC, which refer to different views or angles, and L and R, indicating the left or right breast. We kept these labels intact to preserve important information. Like with the lung cancer dataset, the images were converted into NumPy arrays for easier processing. We only included cases that had masses, leaving out those with calcifications, to keep things consistent with our lung cancer data. After creating both datasets, we normalized the images to ensure that the pixel values (which represent different colors or grayscale levels) were within the same range across all images. This step was important to improve the accuracy of the U-Net algorithm used for segmentation, as consistent input data leads to better model performance.

We also checked the CBIS-DDSM images for size consistency. Unlike the lung cancer images, not all of the breast cancer images were the same size. To fix this, we resized all images to a uniform dimension. When resizing, we added a black background to keep the original shapes of the lungs, breasts, and nodules intact. Images that were smaller than the target size were scaled up, with the original image centered and the surrounding area filled with black, so we could maintain the accuracy of nodule segmentation. To segment the lung and breast regions, we used simple morphological operations to generate Region of Interest (ROI) masks. These masks focused the analysis on the key areas where nodules might be found. By zeroing in on these specific regions, we helped the segmentation algorithms perform more accurately, making it easier for the models to identify and classify the nodules.

6. U-Net architecture

The U-Net architecture is composed of two main parts: a contracting path on the left and an expansive path on the right. The contracting path follows the structure

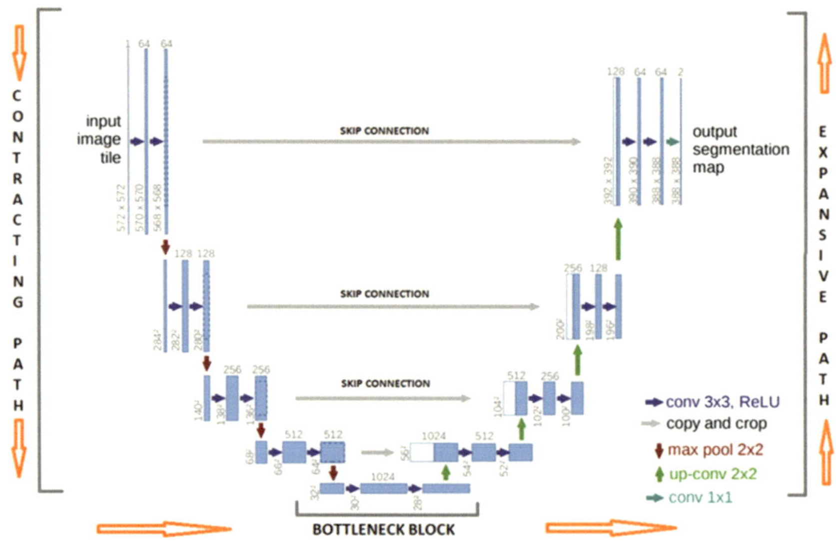

Figure 2.
U-Net architecture.

of a typical convolutional network, where two 3 × 3 convolutions (without padding) are applied repeatedly, each followed by a rectified linear unit (ReLU). After each convolution, a 2 × 2 max-pooling operation with a stride of two is used to downsample the image. With each downsampling step, the number of feature channels is doubled. On the expansive path, the process is reversed. Each step starts with upsampling the feature map, followed by a 2 × 2 convolution (also known as "up-convolution") that reduces the number of feature channels by half. The upsampled feature map is then concatenated with the corresponding cropped feature map from the contracting path. This is followed by two 3 × 3 convolutions, each accompanied by a ReLU activation. Cropping is necessary at this stage to account for the loss of border pixels that occurs during each convolution. At the final layer, a 1 × 1 convolution is applied to convert the 64-component feature vectors into the desired number of output classes. Altogether, the network contains 23 convolutional layers. To ensure that the output segmentation map can be tiled correctly, the input tile size must be chosen so that every 2 × 2 max-pooling operation is performed on a layer with even x and y dimensions. The architecture of the U-Net that is going to be used in this paper is shown in **Figure 2**.

7. Implementation

In this phase, we combined and processed the datasets to prepare them for training and testing the model. The approach involved selecting random samples from both datasets and organizing the data for model input.

We began by loading both the LIDC-IDRI and CBIS-DDSM datasets. From each dataset, we randomly selected 500 patient IDs. The LIDC-IDRI dataset includes multiple images per patient, so for each selected patient, we randomly chose one image. In contrast, the CBIS-DDSM dataset has only one image per patient, so each patient's single image was used. This selection process resulted in a total of 1000 random images—500 from each dataset. Each image was paired with its corresponding mask (for segmentation) and label (for classification, where labels are binary: 0 for benign and 1 for malignant). This method ensured equal representation of both lung and breast cancer cases in the combined dataset. After selecting the images, we organized them into three arrays: images, masks, and labels. Each array was aligned so that the corresponding mask and label for an image were at the same index in their respective arrays. This setup allowed us to maintain the relationship between each image, its mask, and its classification label. To avoid any bias and ensure that the data was properly randomized, all three arrays—images, masks, and labels—were shuffled using the same randomization process. This ensured that the consistency between images, masks, and labels was preserved even after shuffling. Once the data was randomized, we divided the arrays into training, validation, and test sets. The training set was used to train the model, while the validation set helped fine-tune hyperparameters and prevent overfitting. The test set was used to assess the final model's performance.

7.1 Model training and evaluation

The combined, normalized datasets were then fed into a U-Net model, designed for image segmentation and classification. The U-Net model, known for its effectiveness in medical image segmentation, was particularly suitable for this task. It was configured with mechanisms to prevent overfitting, such as dropout layers, and checkpoints to save the best-performing model during training. During the training process, the model learned to predict binary masks for segmentation (indicating the presence of nodules or masses) and to classify the images as benign or malignant. After training, the model was evaluated on the test set, where it returned the predicted binary masks and classification results, demonstrating its ability to accurately segment and classify both lung and breast cancer cases.

8. Results and discussion

The U-Net model was tested on a combined dataset of lung and breast cancer images, and the results were mixed, highlighting both the potential and the challenges of this approach.

8.1 Segmentation performance

The model's ability to accurately segment cancerous regions was far from ideal. Here is a closer look at the metrics:

- Segmentation accuracy: The model achieved a segmentation accuracy of 0.0, which is as low as it gets. This suggests that the model struggled to correctly identify and segment the cancerous areas within the images, missing the mark completely.

- Mean IoU and dice coefficient: The mean IoU and Dice scores, which measure how well the predicted segmentations overlap with the actual nodule regions, were also extremely low. These scores indicate that the model was not able to effectively distinguish between cancerous nodules and other structures in the images. It seems that the U-Net model got confused by the varying shapes and densities in the lung and breast images, leading to inaccurate segmentations.

8.2 Classification performance

On the classification side, the model performed somewhat better, though it still has room for improvement:

- Classification accuracy: The model correctly classified about 68.5% of the cases. While this is better than random guessing, it is not reliable enough for clinical use.

- Confusion matrix: The confusion matrix shows that the model correctly identified 133 malignant cases, but it also mistakenly labeled 67 benign cases as malignant. This high number of false positives means that the model might cause unnecessary concern by flagging non-cancerous cases as cancer.

- ROC-AUC score: The ROC-AUC score of 0.685 indicates that the model has moderate ability to differentiate between benign and malignant cases. It is decent, but not quite at the level needed for confident predictions in a clinical setting.

- F1 score: The F1 score, which balances precision and recall, was 0.7987. While this suggests the model is reasonably balanced, the high false positive rate still points to significant issues, particularly in distinguishing between benign and malignant cases.

The results are displayed in the **Figure 3**.

These results highlight the difficulties in using a single U-Net model to handle both lung and breast cancer images. The model's poor segmentation performance suggests that it has trouble generalizing across the very different image types—lung CT scans and breast mammograms—each with its own set of complexities. The model seems to misinterpret various structures as nodules or masses, leading to inaccurate segmentations.

On the classification front, while the model does better, it still struggles with a high rate of false positives. This suggests that the model, trained on a mixed dataset, might not be finely tuned to the specific features that differentiate benign from malignant cases in lung and breast images.

Overall, these findings suggest that expecting a single model to excel in segmenting and classifying both lung and breast cancer is overly ambitious. The model's confusion between different types of nodules and masses could be addressed by developing specialized models for each type of cancer or by using more advanced techniques that can better handle the diversity in medical images.

Although U-Net has demonstrated great success in various imaging tasks, its performance can vary significantly depending on the imaging modality. For example, when applied to CT scans, U-Net may struggle to capture subtle differences in tissue density compared to mammography, where structural clarity is often higher. These limitations highlight the challenges of generalizing U-Net across different modalities, and its effectiveness can be reduced when tasked with differentiating between more complex anatomical structures. Understanding these limitations is essential for identifying areas where additional modifications or alternative architectures may be required to improve performance across a broader range of medical images.

If these challenges can be addressed, there is potential to integrate these models into web-based applications for early cancer detection, providing a useful tool for healthcare professionals and improving early diagnosis rates. However, as the current

```
from sklearn.metrics import roc_auc_score, f1_score, confusion_matrix, accuracy_score

segmentation_preds, classification_preds = model.predict(X_test)
classification_preds = (classification_preds > 0.5).astype(np.uint8)
# Calculate metrics for segmentation
segmentation_accuracy = model.evaluate(X_test, y_test_masks, verbose=0)[1]
segmentation_mean_iou = model.evaluate(X_test, y_test_masks, verbose=0)[2]
segmentation_dice = model.evaluate(X_test, y_test_masks, verbose=0)[3]

# Calculate metrics for classification
classification_accuracy = accuracy_score(y_test_labels, classification_preds)
roc_auc = roc_auc_score(y_test_labels, classification_preds)
f1 = f1_score(y_test_labels, classification_preds)
conf_matrix = confusion_matrix(y_test_labels, classification_preds)

print(f"Segmentation Accuracy: {segmentation_accuracy}")
print(f"Segmentation Mean IoU: {segmentation_mean_iou}")
print(f"Segmentation Dice Coefficient: {segmentation_dice}")
print(f"Classification Accuracy: {classification_accuracy}")
print(f"ROC AUC: {roc_auc}")
print(f"F1 Score: {f1}")
print(f"Confusion Matrix:\n{conf_matrix}")

7/7 ──────────────── 40s 6s/step
Segmentation Accuracy: 0.0
Segmentation Mean IoU: 0.0
Segmentation Dice Coefficient: 3.959990668711555e-10
Classification Accuracy: 0.665
ROC AUC: 0.5
F1 Score: 0.7987987987987988
Confusion Matrix:
[[  0  67]
 [  0 133]]
```

Figure 3.
Model implementation results.

results show, there's still significant work to be done before such a system can be considered reliable and effective. To further validate the findings, it would be beneficial to apply cross-validation using external datasets. This ensures that the U-Net model generalizes well beyond the CBIS-DDSM and LIDC-IDRI datasets, increasing confidence in its robustness across different imaging sources. Moreover, expanding the evaluation to include more diverse patient demographics would provide insight into the model's ability to perform consistently across various population groups. Including such diverse datasets can also help address potential biases, ensuring that the model is applicable to real-world scenarios involving patients from various ethnicities, ages, and medical histories.

9. Conclusion

This study explored the use of a U-Net model for segmenting and classifying both lung and breast cancer from a combined dataset. The results show that while the model has some ability to classify cancerous cases, its segmentation performance was poor, particularly when trying to handle both types of cancer at once.

These results suggest that more targeted approaches are needed. Future research could focus on developing separate models for lung and breast cancer or incorporating advanced methods like multi-task learning to improve accuracy. Enhancing the preprocessing steps, using better data augmentation, or incorporating additional clinical data might also help.

Author details

Ajna Fetić* and Nevzudin Buzađija
University of Zenica, Zenica, Bosnia and Herzegovina

*Address all correspondence to: ajna.fetic2018@size.ba

References

[1] Islam MM, Haque MR, Iqbal H, Hasan MM, Hasan M, Kabir MN. Breast Cancer Prediction: A Comparative Study Using Machine Learning Techniques. Singapore: Springer Nature; 2020

[2] Jacobs C, Murphy K, Prokop M, van Ginneken B. Computer-aided detection of pulmonary nodules: A comparative study using the public LIDC/IDRI database. European Radiology (Heidelberg, Germany: Springer). 2016

[3] Al-Shabi M, Boon LL. Lung Nodule Classification Using Deep Local-Global Networks. Heidelberg, Germany: Springer; Sep 2020

[4] Hanliang J, Fuhao S, Fei G. Learning Efficient, Explainable, and Discriminative Representations for Pulmonary Nodule Classification. Arxiv, Cornell University; Jan 2021

[5] Al-Shabi M, Hwee KL. Gated-Dilated Networks for Lung Nodule. Arxiv, Cornell University; 2019

[6] Ray S. A quick review of machine learning algorithm. In: International Conference on Machine Learning, Big Data, Cloud and Parallel Computing. Sep 2019

[7] Al-Shabia M, Shaka K, Tana M. ProCAN: Progressive Growing Channel Attentive Non-Local Network for Lung Nodu Classification. Arxiv, Cornell University; 2020

[8] Al-Shabi M, Shak K. ProCAN: Progressive Growing Channel Attentive Non-Local Network for Lung Nodule Classification. Elsevier; 2021

[9] Wang S, Zhou M. A Multi-View Deep Convolutional Neural Network for Lung Nodule Segmentation. Publmed. 2022

[10] Nasrullah N, Sang J. Automated lung nodule detection and classification using deep learning combined with multiple strategies. Sensors. 2020

Section 3

Innovation, Optimization, and Strategic Management in Healthcare Systems

Chapter 6

Optimizing Healthcare Delivery: Innovations and Economic Strategies in Medical Device Management

Ricardo Marcão, Stéphanie Monteiro, Vasco Santos, Filipa Martinho, Maria José Sousa, António Dionísio and Pedro Ramos

Abstract

The global healthcare industry is currently undergoing a significant technological transformation, with the digitization of medical devices and systems at the forefront of revolutionizing healthcare delivery. This evolution aims to make healthcare more accessible, effective, and cost-efficient, necessitating healthcare providers to strategically evaluate and embrace new technologies along with procedural changes. This paper offers a comprehensive analysis of optimizing healthcare delivery, focusing on the crucial aspect of medical device management throughout its lifecycle. It distills academic research into practical guidance for healthcare providers and medical professionals, emphasizing the importance of patient-centric care, risk management, regulatory compliance, safety standards, and vigilant post-market surveillance. The lifecycle approach, covering development, testing, manufacturing, marketing, supply chain, device management, and eventual retirement, underscores the importance of safe and effective medical device management. It demonstrates that modern healthcare's reliance on medical technology demands thoughtful and robust management processes for the efficient, safe, and innovative utilization of devices, benefiting both patients and organizations. Through discussions on research and case studies, the paper outlines a strategic process for achieving an economical and patient-centric healthcare delivery system, addressing the impact of healthcare costs on the utilization of medical devices and promoting studies on advancing patient-oriented healthcare systems.

Keywords: medical device management, healthcare delivery, technological innovation, patient-centric care, regulatory compliance, predictive maintenance, economic strategies, telemedicine

1. Introduction

The healthcare industry is at a pivotal moment, driven by the rapid pace of technological advancements and the increasing demand for patient-centered care. Medical devices, from basic instruments to advanced digital systems, have become critical in enhancing diagnostics, treatment, and patient management. These devices not only improve clinical outcomes but also facilitate access to care across diverse populations, particularly as healthcare moves toward more personalized and preventive models [1]. However, effectively managing these devices throughout their lifecycle—from procurement and maintenance to compliance with regulatory standards and eventual decommissioning—presents several challenges for healthcare providers [2]. These challenges are further compounded by rising healthcare costs, the need for technological integration, and the imperative of ensuring safety and efficacy in device management [3].

The overarching goal of this chapter is to provide a comprehensive strategic analysis of medical device management as a key element in optimizing healthcare delivery. This chapter will focus on the integration of technological innovations, economic strategies, and a forward-looking approach to future trends. Technological advancements such as artificial intelligence (AI), the Internet of Things (IoT), and predictive maintenance systems are transforming not only how medical devices are managed but also how healthcare is delivered. These technologies offer opportunities to reduce costs, enhance operational efficiency, and improve patient outcomes by enabling more precise, real-time, data-driven decision-making [1]. The economic implications of these advancements are equally significant, as healthcare providers must navigate the complexities of managing the high costs associated with device acquisition, operation, and maintenance while upholding high standards of care [3].

The first section of this chapter will explore the current challenges in medical device management. These challenges include escalating costs, inefficient resource utilization, limited access to advanced medical technologies, and the pressing need for regulatory compliance. The healthcare sector, in particular, faces significant hurdles in maintaining up-to-date inventories of medical devices, ensuring proper maintenance, and adhering to the regulatory requirements that govern device safety and efficacy [2]. Additionally, global disparities in access to essential medical devices, particularly in developing countries, further complicate efforts to optimize healthcare delivery [3]. The discussion will also highlight how the COVID-19 pandemic exposed vulnerabilities in global healthcare systems, particularly in medical device supply chains, amplifying the need for more resilient and innovative device management strategies [4].

Following the examination of these challenges, the chapter will present solutions through detailed case studies that demonstrate both successful and unsuccessful implementations of medical device management strategies. These case studies will offer real-world insights into how healthcare organizations have addressed operational inefficiencies, improved device utilization, and navigated regulatory barriers. For instance, the use of centralized inventory management systems has been shown to reduce device downtime and improve resource allocation in several leading healthcare institutions. Conversely, examples of failed initiatives will provide valuable lessons on the importance of stakeholder engagement, transparent communication, and the need for continuous monitoring and evaluation of medical device management systems [3].

The chapter will also delve into the critical regulatory considerations governing the medical device industry, emphasizing the importance of compliance with both national and international standards. These regulations, such as those enforced by the U.S. Food and Drug Administration (FDA) and the European Union's CE marking, ensure the safety, efficacy, and quality of medical devices throughout their lifecycle. The complexity of these regulatory frameworks highlights the need for healthcare providers to remain informed about evolving standards, particularly as new technologies such as AI-driven diagnostic tools and IoT-enabled medical devices enter the market. Additionally, data security and patient privacy, especially concerning connected devices, remain paramount concerns that must be addressed through stringent cybersecurity measures [4].

Lastly, the chapter will examine future trends in medical device management, focusing on the potential of emerging technologies to revolutionize healthcare delivery. AI, IoT, robotics, and 3D printing are just a few of the innovations that promise to enhance device functionality, improve diagnostic accuracy, and allow for the development of personalized medical devices tailored to individual patient needs. Predictive maintenance, enabled by AI and IoT, will continue to play a crucial role in minimizing device downtime, reducing costs, and improving overall care quality [3]. These trends underscore the need for continuous investment in research and development, regulatory innovation, and workforce training to ensure that healthcare providers can fully leverage the potential of these technologies in delivering more efficient and equitable healthcare [4].

By examining these critical areas, this chapter aims to provide healthcare providers, policymakers, and industry stakeholders with actionable insights and strategies to optimize healthcare delivery through more effective and innovative medical device management. The integration of new technologies, economic strategies, and regulatory compliance will not only improve operational efficiency but also enhance patient outcomes, ultimately leading to a more sustainable and patient-centered healthcare system [4].

2. Current challenges in healthcare delivery

The landscape of healthcare delivery is fraught with challenges, notably heightened by the expansion of international activities in the medical device industry. This globalization has prompted a push toward more efficient production processes, stringent safety and quality controls by government agencies, and cost containment strategies by healthcare providers, with significant attention on optimizing medical device management. The pivotal role of advanced medical technologies, such as intelligent devices including infusion pumps and cardiac pacemakers, has shifted the paradigm from traditional hospital-based care to more economical, patient-centered models. This shift underscores the urgency for rational device management to ensure safety, efficacy, and cost-efficiency, especially as the industry navigates increased market volatility and the evolving demands of both developed and developing economies [5, 6].

Concurrently, healthcare in the United States faces escalating expenditures, with projections suggesting an alarming rise to $6 trillion by 2027, or 19% of GDP. This surge is driven by an aging population, more expensive healthcare services and drugs, the prevalence of chronic diseases, defensive medicine practices, and substantial administrative costs. These factors collectively underscore the necessity for innovative strategies in medical device management to mitigate rising costs and improve system efficiency [7, 8]. Additionally, global access to medical devices remains uneven, highlighting disparities exacerbated by socio-economic and structural barriers, particularly in less developed territories, challenging the globalization of medical device supply and underlining the health inequalities across different populations [9–11]. This increase is multifaceted, rooted in an aging population that demands more medical services, escalating prices of healthcare services and prescription drugs, and the widespread prevalence of chronic diseases, which alone consume about 90% of the country's healthcare spending. Moreover, the fragmented healthcare system exacerbates these costs, leading to inefficient resource use and suboptimal patient outcomes. Defensive medicine practices and the complex administrative burdens of the multi-payer insurance system further inflate these expenditures. Addressing these escalating costs is crucial, necessitating innovative approaches in medical device management to enhance efficiency and curb unnecessary spending [7, 8].

Moreover, inefficiencies in device management within healthcare facilities, such as unclear allocation processes, inadequate stock visibility, and subpar device recall management, compromise the quality of patient care and resource utilization. The absence of streamlined processes and reliance on outdated management methods underscore the critical need for improvement in inventory and asset management systems to enhance care delivery and operational efficiency [12–14].

The global disparity in access to medical devices underscores a critical yet underexplored issue in healthcare delivery. While advancements and the need for innovation in medical device development have garnered attention, particularly in economically advanced regions, this focus does not fully capture the challenges faced by many worldwide. A significant number of healthcare patients, practitioners, and providers across various regions struggle to access essential medical devices. This gap is primarily driven by the growing demand for innovative devices needed to tackle contemporary health issues—a demand influenced heavily by the rising prevalence of chronic conditions in aging populations. These challenges to the globalization of medical device manufacture and supply are further compounded by socio-economic and structural barriers, particularly in less developed areas. These regions often lack the robust healthcare infrastructure and regulatory frameworks essential for the advancement and maintenance of medical devices, thus perpetuating the divide between 'innovator' and 'adopter' countries. This not only diminishes the attractiveness of different healthcare markets to medical device manufacturers but also exacerbates health inequalities across patient populations [9–11].

Inefficient device management in hospitals, characterized by unclear allocation processes and the absence of standardized procedures, results in operational inefficiencies and increased costs. Lack of clear priorities and protocols often leaves high-demand services without essential devices, complicating emergency responses [12]. The management of device recalls suffers due to inadequate monitoring, leading to incomplete maintenance tasks and poorly regulated inventory, heightening the risk of shortages or surplus stock [13]. Additionally, the reorder process is hampered by inefficiencies such as outdated automation and lack of digitalization, further strained by improper handling of expired items, affecting inventory accuracy and storage conditions [14]. The absence of an effective asset management system and data-driven optimization exacerbates these challenges, with staff wasting time on device location and maintenance scheduling, thereby impacting patient care and device utilization.

3. Importance of optimizing healthcare delivery

Optimizing healthcare delivery is increasingly recognized as a crucial step toward enhancing patient care and addressing the rising costs associated with healthcare services. This optimization is not only beneficial for achieving commercial and financial success but is fundamentally about improving patient outcomes. Effective healthcare delivery focuses on ensuring patient care takes precedence, facilitated by technological advancements and the strategic management of healthcare resources. Interoperability and seamless access to electronic health records (EHRs) significantly enhance care delivery, improving both effectiveness and efficiency across various healthcare settings. Despite challenges, such as delays in compliance and legislative pushbacks, the integration of EHRs is vital for enabling patient-centered outcomes. The drive toward optimized healthcare involves a comprehensive approach that aligns patient experiences with the expertise of medical staff and effective protocol implementation. Highlighting the complexity of this endeavor, there are substantial positive impacts achievable through meticulous planning and execution. Furthermore, leveraging advanced data analytics, including probabilistic linkage and data mining, fosters quality improvement and integrates patient records across the healthcare spectrum [15].

Enhancing patient care stands at the forefront of optimizing healthcare delivery, underlining the essence of patient-centric approaches within the healthcare system. A pivotal element in this optimization is the reduction of medical errors, a concern highlighted across various studies. For instance, research indicates that suboptimal equipment management practices significantly contribute to medical errors, with estimates suggesting that medical device issues lead to approximately 155,000 errors and 1300 fatalities annually [16]. The integration of advanced technologies, such as barcoding systems in surgical supply chains, demonstrates a tangible reduction in human errors, ensuring the correct utilization of medical equipment for each procedure and enhancing patient safety. Furthermore, intelligent systems capable of diagnosing issues and predicting equipment failures mark a substantial advancement in medical device management, moving toward a model of 'preventive maintenance' over traditional methods. This shift not only maximizes equipment uptime but also significantly improves clinical care quality. The deployment of such technologies since their introduction by diagnostic imaging centers reflects the critical role of innovative medical device management in bolstering healthcare delivery and patient care outcomes [17, 18].

Reducing healthcare costs while enhancing the quality of care is a pivotal goal for healthcare organizations. They aim to utilize accurate data on device use and inventory to pinpoint and eliminate inefficiencies in medical device management [19]. By cutting down on unnecessary activities and equipment redundancy, organizations can boost resource efficiency and lower overall healthcare costs without compromising care delivery or quality. This strategy emphasizes the significance of integrating technology and information advancements to enhance patient care and operational productivity [20]. However, these innovations bring their own costs, including the expense of adopting new "smart" devices equipped with advanced diagnostic and treatment capabilities. Such expenses necessitate careful planning to ensure that investments in new technology genuinely contribute to cost reduction and enhanced patient care outcomes. Engaging leaders and forming a multidisciplinary team early in the cost reduction process is critical for ensuring clinical relevance and comprehensive cost-saving strategies [21].

The evolution of medical technology, while offering significant improvements in healthcare delivery, also requires strategic management to balance innovation with cost-effectiveness. The challenge of technology obsolescence, where new models quickly replace the old, needs a planned increase in the utilization of newer models to avoid productivity dilution [22]. Furthermore, other authors emphasize the importance of leveraging new technologies for strategic cost reduction, suggesting that meaningful use and evidence-based treatment practices can offer unique value opportunities in medical device management [23, 24].

4. Innovations in medical device management

The landscape of medical device management is continually evolving, marked by significant innovations that aim to enhance healthcare delivery and patient outcomes. These advancements range from the adoption of electronic health records (EHRs) and telemedicine to the integration of the Internet of Things (IoT) in healthcare practices. EHRs, or electronic medical records (EMRs), serve as digital versions of patients' traditional paper charts, offering a comprehensive record that supports treatment and care delivery across various healthcare settings. This digital transition not only facilitates global collaboration among healthcare providers but also reduces unnecessary patient testing and medication exposure, thereby improving patient safety and reducing costs. The proliferation of EHRs, propelled by the Health Information Technology for Economic and Clinical Health (HITECH) Act in the USA, signifies a critical shift toward digital healthcare systems that prioritize efficiency and patient privacy [25–27].

Telemedicine has emerged as a pivotal innovation, extending healthcare's reach to remote and underserved populations while enhancing the patient-doctor relationship through cost-effective and accessible services. Supported by over 200 research articles, telemedicine's effectiveness in managing chronic illnesses is increasingly acknowledged worldwide, fostering its adoption and global diffusion [28, 29].

Remote patient monitoring (RPM) technologies represent another significant stride in healthcare innovation, enabling real-time health data transmission from patients to healthcare providers. This technology empowers patients to actively participate in their health management, offering a blend of privacy, security, and convenience. RPM is anticipated to redefine medical practice by fostering closer patient-provider connections and facilitating home-based patient care and rehabilitation, supported by federal initiatives like the Centers for Medicare & Medicaid Services Chronic Care Management program [30, 31].

Artificial Intelligence (AI) in medical device management is revolutionizing healthcare by enhancing device efficiency and patient care outcomes. AI technologies facilitate predictive maintenance, remote monitoring, and energy optimization of medical devices, offering a smarter approach to healthcare service delivery. Despite the potential benefits, the full realization of AI in healthcare faces challenges due to the industry's regulated nature and the hesitance of users to adopt new technologies [32, 33].

Predictive maintenance systems stand out as a cost-effective solution to medical device management, shifting away from traditional maintenance strategies to a more data-driven approach. This method improves device reliability, optimizes performance, and extends the useful life of medical equipment, aligning with the healthcare industry's goal of enhancing efficiency and reducing costs [34].

Together, these innovations in medical device management underscore a transformative period in healthcare, driving toward a future where technology and data analytics play central roles in delivering patient-centered, efficient, and effective care.

5. Economic strategies for medical device management

Economic strategies in medical device management are pivotal for healthcare providers to navigate the complexities and high costs associated with the deployment, operation, and maintenance of these critical tools. The transition from reactive to proactive management strategies, facilitated by advancements in technology and the Internet of Things (IoT), offers a promising avenue for reducing the economic burden associated with medical devices.

Value-based purchasing emerges as a strategic approach encouraging healthcare providers to select medical devices not only based on cost-effectiveness but also on their long-term value in patient care and operational efficiency. This method necessitates a comprehensive understanding of device lifecycle costs, demanding an initial investment and potentially additional staff training to achieve optimal outcomes. The emphasis on long-term patient care impact and efficiency aligns with a broader industry shift toward value-oriented healthcare delivery [35, 36].

Equipment standardization represents another key strategy, advocating for uniformity in device selection to streamline operations, reduce training requirements, and minimize maintenance complexity. Despite its initial perceived limitations, standardization can lead to significant long-term benefits, including cost savings through bulk purchasing agreements and improved operational efficiency [37].

Asset tracking and utilization, facilitated by digital systems, provides real-time insights into device location and usage patterns. This strategy enhances the ability to allocate resources effectively, ensuring devices are utilized where they are most needed and supporting strategic planning for future purchases based on actual usage data. Such granular visibility into device utilization patterns is instrumental in optimizing patient care delivery and operational efficiency within healthcare settings [38].

These economic strategies, encompassing maintenance optimization, lifecycle cost analysis, and strategic device placement, require careful consideration of the unique needs of patient populations and healthcare providers. They must be tailored to fit within the broader regulatory, financial, and technological landscape of the healthcare industry. Successful implementation hinges on the adaptability of healthcare providers to these innovative strategies, promising significant benefits in terms of cost savings, improved patient outcomes, and enhanced operational efficiency [39].

The exploration of these strategies within this chapter underscores their potential to transform medical device management in healthcare. By adopting a comprehensive, forward-thinking approach to purchasing, standardization, and asset tracking, healthcare organizations can navigate the challenges of high costs and complex logistics associated with medical devices, moving toward a more efficient, patient-centered, and economically sustainable model of care delivery.

6. Integration of technology in healthcare delivery

The integration of technology into healthcare delivery heralds a pivotal transformation toward a system that is more efficient, accessible, and tailored to individual needs. The inception of this transformation began with the digitalization of healthcare facilities, aiming for a "paperless" environment. This foundational shift made possible the development and integration of advanced technologies such as artificial intelligence (AI), virtual clinics, and natural language processing (NLP) into the healthcare system. These technologies have significantly enhanced the direct interaction between patients and care providers through the adoption of standardized health informatics and telemedicine systems.

Telemedicine and remote patient monitoring stand out as key examples of how technology can extend healthcare beyond traditional settings. Leveraging rapid internet access and advancements in technology, these systems enable continuous, real-time monitoring of patients through wearable and portable devices. This has facilitated care delivery in a variety of settings, thereby improving access to expert healthcare advice and enhancing the personalization of healthcare services. The evolution toward modern maintenance strategies for medical devices, shifting from reactive to predictive maintenance enabled by AI and the Industrial Internet of Things (IIoT), exemplifies this trend. Predictive maintenance systems anticipate the needs for device maintenance, significantly reducing downtime and associated costs, which in turn improves the quality of patient care [40].

Electronic Health Records (EHRs) have become a cornerstone in this technological integration, providing a longitudinal digital record of patient medical histories that can be shared across various healthcare settings. The widespread adoption of EHRs, spurred by the Health Information Technology for Economic and Clinical Health (HITECH) Act, emphasizes the critical role of digital records in improving care coordination, enhancing diagnosis and patient outcomes, and ensuring patient safety. This transition toward digitalization aligns with broader trends toward optimizing healthcare delivery through innovations in medical device management, highlighting the emphasis on efficiency, data analytics, and a patient-centered approach to care [41].

Furthermore, telemedicine epitomizes the technological integration in healthcare by reducing medical costs and enhancing the accessibility of services. It provides a cost-effective model for healthcare delivery by reducing patient travel and facility capital investment, thereby improving service quality and flexibility. This modality is particularly beneficial in outpatient and home settings, indicating a shift toward more accessible and efficient healthcare delivery modalities [42].

The Internet of Things (IoT) introduces a network of interconnected devices capable of collecting and exchanging data, ranging from wearable technologies to implantable medical devices. The potential benefits of IoT in healthcare are vast, offering improved patient monitoring, preventive care, and insights into disease progression, which can contribute to a shift toward a more personalized healthcare system. Nonetheless, the realization of IoT's full potential in healthcare necessitates addressing challenges related to data management, privacy, and regulation of IoT devices [43–47].

As these technologies continue to advance and become further integrated into healthcare delivery, the vision of a healthcare system that is more efficient, accessible, and personalized becomes increasingly attainable. The ongoing innovation and adoption of digital health technologies, underpinned by advancements in AI, telemedicine, and IoT, promise a future where healthcare delivery is optimized to meet the varied needs of patients and providers alike.

7. Regulatory considerations in medical device management

Navigating the regulatory landscape is crucial for ensuring the safety, efficacy, and quality of medical devices throughout their lifecycle. This complex environment encompasses a variety of national and international standards and regulations that manufacturers and healthcare organizations must adhere to, from product development to post-market surveillance [48].

The regulatory framework varies significantly across different jurisdictions, highlighting the importance of a nuanced approach to compliance. For instance, in the United States, the Food and Drug Administration (FDA) sets forth rigorous guidelines for device registration, approval, and quality controls [49]. Similarly, the European Union requires medical devices to attain CE Marking as a demonstration of safety and performance, underlining the global diversity in regulatory requirements [50].

Intelligent medical devices, which are increasingly prevalent in healthcare for applications like remote monitoring, must not only meet safety and performance standards but also ensure data accuracy and protection. This necessitates robust security measures and standards to safeguard sensitive patient information and justify clinical decisions based on device-generated data [51].

Manufacturers must also navigate the challenges posed by technological advancements, ensuring that devices on the market remain compliant through active post-market surveillance. This involves systematic data collection and analysis to monitor product performance and safety and to take corrective actions if necessary [52].

Compliance with FDA regulations, a cornerstone of medical device management in the U.S., involves several key steps, including premarket notification (510(k)), premarket approval (PMA), and the De Novo classification for novel devices. These processes underscore the complexity of bringing medical devices to market and the critical role of compliance in safeguarding patient health [53].

Data security and privacy are paramount, with regulations like the Health Insurance Portability and Accountability Act (HIPAA) in the U.S. setting standards for the protection of personal health information. The increasing threat of cyber-attacks on the medical device sector amplifies the need for comprehensive cybersecurity risk management strategies to protect patient data [54].

Quality assurance (QA) and quality control (QC) are integral to maintaining high standards in medical device management, focusing on process and product quality to ensure that devices are safe, effective, and fit for their intended use. Adherence to global medical device regulations and standards, such as those set by the FDA and CE marking, is essential for patient safety, regulatory compliance, market access, and organizational reputation in the healthcare sector [55–58].

Overall, the regulatory considerations in medical device management encompass a broad spectrum of activities and standards, from compliance and data security to quality assurance and control, underscoring the critical role of regulation in ensuring the safe and effective use of medical devices in healthcare.

8. Case studies in optimizing healthcare delivery

Case studies are valuable tools for understanding how theoretical strategies can be successfully or unsuccessfully applied in real-world scenarios. This section presents two contrasting case studies that highlight both the benefits and challenges of medical device management. Through these examples, we explore how different healthcare organizations implemented centralized and decentralized device management strategies, offering insights into best practices and lessons learned.

8.1 Case study: MacNeal Hospital's centralized device management system

MacNeal Hospital, a prominent healthcare facility in Illinois, undertook a comprehensive overhaul of its medical device management strategy in response to operational inefficiencies caused by a decentralized system. Previously, the hospital struggled with maintaining accurate records of device usage, timely maintenance, and equipment availability, which led to prolonged downtime for essential medical devices. These issues negatively impacted patient care and led to increased operational costs.

Recognizing the need for improvement, MacNeal Hospital implemented a centralized inventory and device management system. This new system allowed for real-time tracking of devices, predictive maintenance scheduling, and a streamlined process for handling device recalls. By centralizing their management system, the hospital could reduce the downtime of critical equipment, ensuring that devices were available when needed for patient care. Additionally, the centralized system enabled better compliance with regulatory standards, as it allowed for more consistent monitoring of device performance and documentation of maintenance activities.

The results were significant: MacNeal Hospital saw a 20% reduction in equipment downtime, a 30% increase in operational efficiency, and an improvement in patient outcomes due to the timely availability of devices for diagnostics and treatment. Furthermore, the hospital experienced cost savings by reducing unnecessary purchases and eliminating equipment redundancies. The successful implementation of the centralized device management system at MacNeal Hospital serves as a model for other healthcare institutions looking to improve their operations through better resource allocation and management practices [59].

8.2 Case study: Health information technology project failure

Not all attempts to optimize medical device management have yielded positive outcomes. A notable example of this is the failure of a large-scale health information technology project at a multi-hospital system in the United States. The project aimed to integrate medical device management with the hospitals' electronic health record (EHR) system to improve interoperability, device tracking, and patient data accuracy.

The integration was intended to streamline workflows, enabling healthcare staff to monitor devices and patient data from a single interface. However, the project encountered significant issues due to a lack of stakeholder engagement, poor communication between IT and clinical teams, and inadequate training for end-users. These factors led to widespread user resistance, and the system was not fully adopted by the healthcare staff.

The failure to fully implement the system resulted in continued inefficiencies, including inaccurate device tracking, delayed maintenance, and ongoing compliance issues with regulatory standards. Furthermore, the lack of proper training contributed to staff frustration and misuse of the new system, exacerbating existing problems rather than solving them.

One of the key lessons learned from this failure was the importance of involving all stakeholders—from IT professionals to clinical staff—in the planning and implementation phases of technology integration projects. Additionally, the project highlighted the need for comprehensive training programs to ensure that healthcare staff can fully utilize new systems and workflows. This case study underscores the fact that even with advanced technology, successful implementation requires clear communication, effective training, and ongoing support to ensure adoption and long-term sustainability [60].

These two case studies illustrate the challenges and potential benefits of medical device management strategies. MacNeal Hospital's success with centralized management underscores the importance of resource visibility, predictive maintenance, and real-time monitoring in achieving operational efficiency and cost savings. On the other hand, the failure of the health information technology integration project highlights the dangers of underestimating the human factors involved in implementing new technologies.

Healthcare providers aiming to optimize device management must carefully plan their strategies, ensuring that stakeholders are engaged, training programs are robust, and technological solutions are tailored to the needs of both staff and patients. By learning from both successful and failed implementations, healthcare organizations can better navigate the complexities of medical device management and improve their overall healthcare delivery systems [61–63].

9. Future trends in medical device management

The healthcare landscape is rapidly evolving, with emerging technologies poised to transform the management of medical devices and revolutionize healthcare delivery. In the coming decade, innovations such as artificial intelligence (AI), the Internet of Things (IoT), robotics, and personalized medical devices are expected to significantly improve operational efficiency, reduce costs, and enhance patient outcomes. These advancements promise to address some of the most pressing challenges in healthcare, including the rising costs of care, the need for real-time monitoring, and the growing demand for personalized treatment options.

9.1 Artificial intelligence (AI) in medical device management

Artificial intelligence is playing an increasingly vital role in medical device management, offering the potential to drastically improve efficiency and safety through predictive maintenance systems. By analyzing vast amounts of data from connected medical devices, AI algorithms can predict when a device is likely to fail, allowing for preemptive maintenance and reducing the risk of downtime. This shift from reactive to predictive maintenance has been shown to reduce costs, increase device uptime, and improve overall patient care quality [64].

AI also enhances decision-making in healthcare by providing insights into device usage patterns and operational inefficiencies. For instance, AI-powered analytics can help healthcare providers optimize the allocation of devices, ensuring that critical equipment is available where and when it is needed. Additionally, AI can be integrated into diagnostic tools, enabling devices to assist clinicians in detecting diseases earlier and more accurately, leading to better patient outcomes. Over the next decade, AI's role in optimizing device management is expected to expand as healthcare organizations increasingly adopt AI-driven solutions to improve both the operational and clinical aspects of device usage [64].

9.2 Internet of things (IoT) and real-time device monitoring

The Internet of Things is another technology that will significantly impact medical device management, particularly in the areas of real-time monitoring and data exchange. IoT-enabled devices are capable of transmitting real-time data to healthcare providers, allowing for continuous monitoring of patient health. This technology enables clinicians to track vital signs, device performance, and treatment progress remotely, leading to more proactive interventions and reducing the need for in-person hospital visits [65].

In the context of medical device management, IoT devices can streamline workflows by automating the process of tracking device status, usage, and location. Hospitals and clinics can use this information to ensure that devices are maintained properly, reducing the risk of malfunctions and extending the lifespan of the equipment. IoT also offers the potential for enhanced interoperability between devices, improving communication across healthcare systems, and facilitating the integration of data from multiple sources. As IoT adoption grows, healthcare providers will be able to harness this data to optimize care delivery, reduce operational costs, and improve patient outcomes [65].

9.3 Robotics in healthcare and medical device management

Robotics is becoming increasingly prevalent in healthcare, particularly in surgical settings where robotic systems assist surgeons in performing complex procedures with greater precision and control. In medical device management, robotics plays a key role in automating routine tasks, such as sterilization, device assembly, and even distribution within hospitals. The use of robotic systems can reduce human error, improve efficiency, and enhance the quality of care by ensuring that medical devices are properly maintained and available when needed [66].

Additionally, telerobotics—where surgeons can perform operations remotely using robotic systems—opens up new possibilities for healthcare delivery in rural and underserved areas. This technology allows highly skilled specialists to perform procedures without being physically present, reducing geographical barriers to quality care. As robotics continues to advance, its integration into both clinical and operational aspects of medical device management will significantly enhance healthcare delivery [66].

9.4 Impact of emerging technologies on healthcare delivery

The integration of AI, IoT, robotics, and personalized devices will fundamentally alter the healthcare landscape in the coming years. These technologies will enable healthcare providers to deliver more efficient, cost-effective, and patient-centered care. Predictive maintenance systems powered by AI will reduce device downtime, lower operational costs, and improve patient safety by ensuring that critical devices are always functional when needed. IoT-enabled devices will allow for continuous patient monitoring, facilitating early detection of health issues and enabling timely interventions that improve patient outcomes [65].

Robotics will automate routine tasks and enable remote surgeries, reducing the burden on healthcare workers and expanding access to specialized care in underserved regions. Personalized devices created through 3D printing will transform the way healthcare providers meet the unique needs of patients, offering better customization and improving the quality of life for those with complex medical conditions. As these technologies continue to develop and become more widely adopted, they will redefine how healthcare is delivered, ultimately leading to a more efficient, accessible, and patient-centric system [66, 67].

10. Conclusion

Optimizing healthcare delivery is essential for enhancing patient care and addressing rising healthcare costs. This chapter has highlighted the critical role of strategic

DOI: http://dx.doi.org/10.5772/intechopen.1007476

medical device management, facilitated by technological advancements such as artificial intelligence (AI), the Internet of Things (IoT), and predictive maintenance systems. By adopting these innovations, healthcare organizations can improve both operational efficiency and patient outcomes.

A primary focus in healthcare delivery optimization is enhancing patient care through better management processes. The integration of AI and IoT into medical device management offers significant advantages, such as predictive maintenance systems, which can reduce device downtime and improve safety. Predictive analytics also enable proactive identification of device failures, ensuring critical equipment remains operational and improving clinical care quality. The incorporation of these advanced technologies fosters a safer and more efficient healthcare environment [16–18].

Reducing medical errors is another crucial goal. Research shows that improper device management contributes significantly to medical errors, with estimates suggesting that device-related issues lead to approximately 155,000 errors and 1300 fatalities annually. Implementing barcoding systems in surgical supply chains and healthcare operations can minimize human error, ensuring proper device usage and improving patient safety. By adopting preventive maintenance strategies, healthcare providers can further reduce equipment failures and maximize device uptime, directly enhancing patient care [16–18].

From an economic perspective, optimizing device management not only improves safety but also leads to substantial cost savings. Investing in value-based purchasing, which considers the long-term value of devices in addition to their upfront costs, is key to sustainable healthcare delivery. Moreover, standardizing medical devices across healthcare facilities can streamline operations, reduce training times, and simplify maintenance, all while minimizing operational expenses. Effective asset tracking systems provide real-time data on device usage, improving resource allocation and reducing unnecessary expenditures [19].

Looking ahead, emerging technologies such as robotics, 3D printing, and advanced data analytics will further reshape the healthcare landscape. These innovations promise to personalize patient care, improve diagnostic and therapeutic precision, and make healthcare more accessible and cost-effective. However, their successful adoption will require careful planning, regulatory compliance, and investment in workforce training and infrastructure [66].

Recommendations: To fully realize the benefits of optimized healthcare delivery, healthcare providers and policymakers should focus on:

1. *Predictive maintenance*: Invest in AI-driven predictive maintenance systems to reduce equipment downtime and ensure critical devices are operational when needed.

2. *Regulatory compliance*: Prioritize compliance with national and international regulatory standards to ensure safety and quality in device management.

3. *Technology integration*: Embrace AI, IoT, and robotics to improve device management and healthcare delivery. These technologies not only enhance operational efficiency but also contribute to better patient outcomes.

4. *Resource allocation*: Implement data-driven asset tracking and resource allocation strategies to optimize device usage and reduce operational inefficiencies.

5. *Standardization and interoperability*: Standardize medical devices and ensure seamless integration of electronic health records (EHRs) to improve care coordination and patient outcomes [67].

In conclusion, optimizing healthcare delivery requires a multifaceted approach that integrates advanced technologies, patient-centered care, and strategic resource management. By focusing on predictive maintenance, regulatory compliance, and technological innovation, healthcare providers can enhance patient safety, reduce costs, and improve overall system efficiency. This chapter underscores the importance of addressing these areas to ensure that healthcare delivery is not only optimized but also sustainable and equitable [67].

Acknowledgements

The authors acknowledge the use of Grammarly Writing Assistant for language polishing of the manuscript.

Author details

Ricardo Marcão[1*], Stéphanie Monteiro[2], Vasco Santos[1,3,4], Filipa Martinho[1,5], Maria José Sousa[6], António Dionísio[1] and Pedro Ramos[1]

1 ISLA Santarém, Santarém, Portugal

2 Institute for Systems and Robotics, Instituto Superior Técnico, University of Lisbon, Lisbon, Portugal

3 Institute Polytechnic of Tomar, Portugal

4 GOVCOPP—Research Unit on Governance, Competitiveness and Public Policies, and CiTUR—Centre for Tourism Research, Development and Innovation, Polytechnic of Leiria, Portugal

5 CEPESE, Oporto, Portugal

6 University Institute of Lisbon, Lisbon, Portugal

*Address all correspondence to: ricardo.marcao@islasantarem.pt

References

[1] Raihan A, Begum RA, Said MNM, Pereira JJ. Relationship between economic growth, renewable energy use, technological innovation, and carbon emission toward achieving Malaysia's Paris agreement. Environment Systems and Decisions. 2022;**42**(4):586-607

[2] Harputlugil T, de Wilde P. The interaction between humans and buildings for energy efficiency: A critical review. Energy Research & Social Science. 1 Jan 2021;**71**:101828

[3] Narula S, Kumar A, Prakash S, Dwivedy M, Puppala H, Talwar V. Modelling and analysis of challenges for industry 4.0 implementation in medical device industry to post COVID-19 scenario. International Journal of Supply and Operations Management. 2023;**10**(2):117-135

[4] Romagnoli A, Ferrara F, Langella R, Zovi A. Healthcare systems and artificial intelligence: Focus on challenges and the international regulatory framework. Pharmaceutical Research. Apr 2024;**41**(4):721-730

[5] Jakovljevic M, Liu Y, Cerda A, Simonyan M, Correia T, Mariita RM, et al. The global south political economy of health financing and spending landscape-history and presence. Journal of Medical Economics. 2021;**24**(Suppl. 1):25-33

[6] Karabag SF. An unprecedented global crisis! The global, regional, national, political, economic and commercial impact of the coronavirus pandemic. Journal of Applied Economics and Business Research. 2020;**10**(1):1-6

[7] Mantus J, Pang G, Warshawsky MJ. A Unified Long-Run Macroeconomic Projection of Health Care Spending, the Federal Budget, and Benefit Programs in the US. AEI Economics Working Paper; 2023

[8] Hartman M, Martin AB, Whittle L, Catlin A, National Health Expenditure Accounts Team. National Health Care Spending in 2022: Growth similar to prepandemic rates. Health Affairs. 2024;**43**(1):6-17

[9] Pradhan B, Bhattacharyya S, Pal K. IoT-based applications in healthcare devices. Journal of Healthcare Engineering. 2021;**2021**(1):6632599

[10] Chauhan V, Galwankar S, Arquilla B, Garg M, Di Somma S, El-Menyar A, et al. Novel coronavirus (COVID-19): Leveraging telemedicine to optimize care while minimizing exposures and viral transmission. Journal of Emergencies, Trauma, and Shock. 2020;**13**(1):20-24

[11] Ratta P, Kaur A, Sharma S, Shabaz M, Dhiman G. Application of blockchain and internet of things in healthcare and medical sector: Applications, challenges, and future perspectives. Journal of Food Quality. 2021;**2021**(1):7608296

[12] Shahid J, Ahmad R, Kiani AK, Ahmad T, Saeed S, Almuhaideb AM. Data protection and privacy of the internet of healthcare things (IoHTs). Applied Sciences. 2022;**12**(4):1927

[13] Spieske A, Gebhardt M, Kopyto M, Birkel H. Improving resilience of the healthcare supply chain in a pandemic: Evidence from Europe during the COVID-19 crisis. Journal of Purchasing and Supply Management. 2022;**28**(5):100748

[14] Jiang L, Wu Z, Xu X, Zhan Y, Jin X, Wang L, et al. Opportunities and

challenges of artificial intelligence in the medical field: Current application, emerging problems, and problem-solving strategies. Journal of International Medical Research. 2021;**49**(3):03000605211000157

[15] Omotayo P, Olasupo J, Alegbeleye T. Centralized electronic health records system: A step towards an optimized health care delivery system ahead of possible future outbreaks. International Journal of Intelligent Information Systems. 2021;**10**(1):1-4

[16] Odeh C. SECTION THREE life cycle/ special populations. In: Primary Care for the Physical Therapist: Examination and Triage. Elsevier. 2020:271

[17] Morgan D, Lukong P, Haywood P, Di Paolantonio G. Examining Recent Mortality Trends: The Impact of Demographic Change. Documents de travail de l'OCDE sur la santé. 21 Nov 2023

[18] Furton GL. The pox of politics: Troesken's tradeoff reexamined. Public Choice. Apr 2023;**195**(1):169-191

[19] Avizenna MH. Applying the Apriori algorithm to analyze and optimize medical device inventory management. Journal of Applied Data Sciences. 2022;**3**(4):143-151

[20] Yao L, Shang D, Zhao H, Hu S. Medical equipment comprehensive management system based on cloud computing and internet of things. Journal of Healthcare Engineering. 2021;**2021**(1):6685456

[21] Milenkovic M, Ciric Lalic D, Vujicic M, Pesko I, Savkovic M, Gracanin D. Project portfolio management in telecommunication company: A stage-gate approach for effective portfolio governance. Project Management. 1 Sep 2023;**30**(5):539-553

[22] Song Y, Ghafari Y, Asefnejad A, Toghraie D. An overview of selective laser sintering 3D printing technology for biomedical and sports device applications: Processes, materials, and applications. Optics & Laser Technology. 1 Apr 2024;**171**:110459

[23] Relan M, Kumar M. Impact of Trust and Risk Factors on the Satisfaction Level of the Policy Holder towards Digital Bancassurance Model. International Research Journal of Management Sociology & Humanity (IRJMSH). 2023

[24] Van Devender MS. Risk Assessment Framework for Evaluation of Cybersecurity Threats and Vulnerabilities in Medical Devices. Doctoral dissertation, University of South Alabama. 2023

[25] Pamidimukkala A, Kermanshachi S. Impact of Covid-19 on field and office workforce in construction industry. Project Leadership and Society. 2021;**19**:23

[26] Euchi J. Do drones have a realistic place in a pandemic fight for delivering medical supplies in healthcare systems problems? Chinese Journal of Aeronautics. 1 Feb 2021;**34**(2):182-190

[27] Xu L, Sanders L, Li K, Chow JC. Chatbot for health care and oncology applications using artificial intelligence and machine learning: Systematic review. JMIR Cancer. 29 Nov 2021;**7**(4):e27850

[28] Tang CS. Innovative technology and operations for alleviating poverty through women's economic empowerment. Production and Operations Management. Jan 2022;**31**(1):32-45

[29] Kolade O, Odumuyiwa V, Abolfathi S, Schroder P, Wakunuma K,

Akanmu I, et al. Technology acceptance and readiness of stakeholders for transitioning to a circular plastic economy in Africa. Technological Forecasting and Social Change. 2022;**183**:121954

[30] Rupasinghe HV, Davis A, Kumar SK, Murray B, Zheljazkov VD. Industrial hemp (Cannabis sativa subsp. sativa) as an emerging source for value-added functional food ingredients and nutraceuticals. Molecules. 2020;**25**(18):4078

[31] Voynova R. Legal balance between the human rights protection and the effectiveness of the criminal investigation. International Conference Knowledge-Based Organization. 2020;**26**(2):255-261

[32] Cioeta R, Cossu A, Giovagnoni E, Rigoni M, Muti P. A new platform for post-marketing surveillance and real-world evidence data collection for substance-based medical devices. Frontiers in Drug Safety and Regulation. 2022;**2**:992359

[33] Musamih A, Salah K, Jayaraman R, Yaqoob I, Puthal D, Ellahham S. NFTs in healthcare: Vision, opportunities, and challenges. IEEE Consumer Electronics Magazine. 2022;**12**(4):21-32

[34] Deep A, Rana AC, Yadav M, Marwaha RK. Regulation of medical devices and their clinical trials studies in the USA: An update. Recent Innovations in Chemical Engineering (Formerly Recent Patents on Chemical Engineering). 2020;**13**(2):94-109

[35] Karne S, Alappan R, DTCD D, Klebonis D, Donelson B. Basics about HIPAA for physicians. This Inaugural Issue of JAAPI is Dedicated to the following Legendary Indian Physicians. 2021;**1**(1):51-55

[36] Bhartendu PS, Shukla S, Misra G. Good practices manufacturing in quality. In: Quality Control and Regulatory Aspects for Biologicals: Regulations and Best Practices. CRC Press; 22 Apr 2024

[37] Upadhyay A, Manzoor A, Bhushan B, Tewari S. Indian industries and biologicals. In: Quality Control and Regulatory Aspects for Biologicals: Regulations and Best Practices. CRC Press. 2024:61

[38] Sharma A, Luthra G. Implementing a risk-based approach to quality management system ISO-13485 processes in compliance with EUMDR 2017/745 for medical device industry. Journal of Pharmaceutical Research International. 2023;**35**(13):8-19

[39] Domingo-Lopez DA, Lattanzi G, Schreiber LH, Wallace EJ, Wylie R, O'Sullivan J, et al. Medical devices, smart drug delivery, wearables and technology for the treatment of diabetes mellitus. Advanced Drug Delivery Reviews. 2022;**185**:114280

[40] Ncube BM, Dube A, Ward K. Establishment of the African medicines agency: Progress, challenges and regulatory readiness. Journal of Pharmaceutical Policy and Practice. 2021;**14**(1):29

[41] MacNeill AJ, Hopf H, Khanuja A, Alizamir S, Bilec M, Eckelman MJ, et al. Transforming the medical device industry: Road map to a circular economy: Study examines a medical device industry transformation. Health Affairs. 2020;**39**(12):2088-2097

[42] Blankart CR, Dams F, Penton H, Kalo Z, Zemplenyi A, Shatrov K, et al. Regulatory and HTA early dialogues in medical devices. Health Policy. 2021;**125**(10):1322-1329

[43] Mane SR, Bais SK, Dhane DD. A review: Perform quality control test for container closures and secondary packing material used in cosmetics. International Journal of Pharmacy and Herbal Technology. ISSN. 2583-8962

[44] MacAdam K. Interoperability and the path to comprehensive health information exchange. NYU Annual Survey of American Law. 2020;**76**:743

[45] Donner CF, ZuWallack R, Nici L. The role of telemedicine in extending and enhancing medical management of the patient with chronic obstructive pulmonary disease. Medicina. 18 Jul 2021;**57**(7):726

[46] Hayes CJ, Dawson L, McCoy H, Hernandez M, Andersen J, Ali MM, et al. Utilization of remote patient monitoring within the United States health care system: A scoping review. Telemedicine and e-Health. 2023;**29**(3):384-394

[47] Boikanyo K, Zungeru AM, Sigweni B, Yahya A, Lebekwe C. Remote patient monitoring systems: Applications, architecture, and challenges. Scientific African. 1 Jul 2023;**20**:e01638

[48] Gerke S, Babic B, Evgeniou T, Cohen IG. The need for a system view to regulate artificial intelligence/machine learning-based software as medical device. NPJ digital medicine. 7 Apr 2020;**3**(1):53

[49] Khan ZF, Alotaibi SR. Applications of artificial intelligence and big data analytics in m-health: A healthcare system perspective. Journal of Healthcare Engineering. 2020;**2020**(1):8894694

[50] Kaparthi S, Bumblauskas D. Designing predictive maintenance systems using decision tree-based machine learning techniques. International Journal of Quality & Reliability Management. 2020;**37**(4):659-686

[51] Khan MA. Challenges facing the application of IoT in medicine and healthcare. International Journal of Computations, Information and Manufacturing (IJCIM). Dec 2021;**1**(1)

[52] Krishnamoorthy S, Dua A, Gupta S. Role of emerging technologies in future IoT-driven healthcare 4.0 technologies: A survey, current challenges and future directions. Journal of Ambient Intelligence and Humanized Computing. 2023;**14**(1):361-407

[53] Degerli M, Ozkan YS. Enablers for IoT regarding wearable medical devices to support healthy living: The five facets. In: IoT in Healthcare and Ambient Assisted Living. Singapore: Springer; 2021. pp. 201-222

[54] Casillo M, Cecere L, Colace F, Lorusso A, Santaniello D. Integrating the internet of things (IoT) in SPA medicine: Innovations and challenges in digital wellness. Computers. 2024;**13**(3):67

[55] Naresh VS, Pericherla SS, Murty PSR, Reddi S. Internet of things in healthcare: Architecture, applications, challenges, and solutions. Computer Systems Science & Engineering. 1 Nov 2020;**35**(6)

[56] Kumar K. A location-allocation model for implementing telemedicine services in primary healthcare network. Nonlinear Studies. 1 Oct 2022;**29**

[57] Planchuelo-Gomez Å, Descoteaux M, Larochelle H, Hutter J, Jones DK, Tax CM. Optimisation of quantitative brain diffusion-relaxation MRI acquisition protocols with physics-informed machine learning. Medical Image Analysis. 1 May 2024;**94**:103134

[58] Filchakova O, Dossym D, Ilyas A, Kuanysheva T, Abdizhamil A, Bukasov R. Review of COVID-19 testing and diagnostic methods. Talanta. 2022;**244**:123409

[59] O'Toole JA, Schmid BSNSW, Level PTAE. Florida: Preventing Medical Errors for PTs and PTAs (311) Course Introduction. atrainceu.com

[60] Taylor JK, Ahmed FZ. Clinical pathways guided by remotely monitoring cardiac device data: The future of device heart failure management? Arrhythmia & Electrophysiology Review. 2023;**12**

[61] Kakhi K, Alizadehsani R, Kabir HD, Khosravi A, Nahavandi S, Acharya UR. The internet of medical things and artificial intelligence: Trends, challenges, and opportunities. Biocybernetics and Biomedical Engineering. 2022;**42**(3):749-771

[62] Jin Z, He C, Fu J, Han Q, He Y. Balancing the customization and standardization: Exploration and layout surrounding the regulation of the growing field of 3D-printed medical devices in. Bio-Design and Manufacturing. Jul 2022;**5**(3):580-606

[63] Shaikh Y, Parvati VK, Biradar SR. Role of IoT and bigdata analytics in healthcare for disease prediction. In: 2020 International Conference on Convergence to Digital World-Quo Vadis (ICCDW). IEEE; 18 Feb 2020. pp. 1-6

[64] Lytras DM, Lytra H, Lytras MD. Healthcare in the times of artificial intelligence: Setting a value-based context. In: Artificial Intelligence and Big Data Analytics for Smart Healthcare. Academic Press; 1 Jan 2021. pp. 1-9

[65] Batko K, Ślęzak A. The use of big data analytics in healthcare. Journal of Big Data. 2022;**9**(1):3

[66] Nazir S, Khan S, Khan HU, Ali S, Garcia-Magarino I, Atan RB, et al. A comprehensive analysis of healthcare big data management, analytics and scientific programming. IEEE Access. 2020;**8**:95714-95733

[67] de Mello BH, Rigo SJ, da Costa CA, da Rosa RR, Donida B, Bez MR, et al. Semantic interoperability in health records standards: A systematic literature review. Health and Technology. 2022;**12**(2):255-272